A Methodology for
Business Success

TOTAL EXPERIENCE

DAR ANDREWS

FIRST EDITION

TOTAL EXPERIENCE:
A METHODOLOGY FOR BUSINESS SUCCESS
Published by John Chisum Press
Omaha, Nebraska city of registration
Copyright © 2024 by Dar Andrews. All rights reserved.

This book contains information obtained from authentic and highly regarded sources. Reasonable efforts have been made to publish reliable data and information, but the author and publisher cannot assume responsibility for the validity of all materials or the consequences of their use. The author and publisher have attempted to trace the copyright holders of all material reproduced in this publication and apologize to copyright holders if permission to publish in this form has not been obtained. If any copyright material has not been acknowledged, please write and let us know, so we may rectify it in any future reprint. Except as permitted under United States Copyright Law, no part of this book may be reprinted, reproduced, transmitted, or utilized in any form in any electronic, mechanical, or other means.

ISBN: 979-8-9906534-0-5

BUSINESS & ECONOMICS / MANAGEMENT

Cover and interior design by Rachel Valliere, copyright owned by Dar Andrews
All rights reserved by Dar Andrews and John Chisum Press.

Printed in the United States of America

To Ava and Savannah—
Dream with grandeur, passion, and ethics.
Then, live your life chasing your dreams!

CONTENTS

Introduction: Why It's Critical to Embrace Total Experience Now ix
What Will You Get Out of This Book?. xi
What Is the Structure of This Book? xii
Who Is the Intended Audience for This Book?. xiv

1 What Is Total Experience? 1
Who Else Should You Be Concerned about Besides Customers and Employees?. 4
So, As You Improve Your Total Experience, You Should See an Immediate Business Benefit, Right? 5
Why Is the Total Experience Business Concept New? 6
Is the Total Experience Concept Only Applicable to For-Profit Businesses?. 7
Chapter 1 Summary 8

2 Which Technologies Are Part of Total Experience? 9
Say What? PayPal Had Better Employee Retention Rates?. 11
So, What Did Donahoe Focus On? 11
What Are the Automation Elements of Total Experience? 12
Are the Elements of Total Experience Applications the Same Across All Businesses? 13
Chapter 2 Summary 15

3 Customer Experience. 17
Do All Customer Experience Applications Have the Same Features?. . 24

How Can a Business Owner Achieve Good Technology Value? 25

Chapter 3 Summary 26

4 Employee Experience 27

How Does AT&T Manage All the Diverse Needs of Its Employees? 28

How Does AT&T Manage the Schedules of All Customer Service Employees? 29

Do I Really Need a Workforce Management Solution? 30

At What Point Does the "Poor Man's Spreadsheet" Option Fail and Require a Formal Application? 31

Common Considerations for Purchasing Workforce Management Software 33

Chapter 4 Summary 34

5 Multi-User Experience 37

Is Multi-User Experience Only Associated with Regulatory Concerns? . . 39

Are There Software or Service Applications Dedicated to Multi-User Experience? 40

To RFP or Not to RFP—That Is the Question! 40

Why Would a Formal Proposal Process Not Always Benefit a Small Entity? 44

Should a Small Business Complete a Proposal for Service Instead of an RFP? 44

Does an RFP Always Increase the Odds of a Successful Total Experience Effort? 45

Chapter 5 Summary 47

6 Operational Excellence 49

How Does This Example Depict the Need for Operational Excellence? 52

What Is Needed to Perform Data Analytics in a Repeatable Manner? . . 52

What Makes Power BI and Looker Studio Analytics Better than a Spreadsheet? 53

How Can You Get Started with Operational Excellence? 54

Chapter 6 Summary 55

7 How Do You Get Started in Total Experience? 57

   Series One: 58

   Series Two: 58

   Prioritize Your Pain Points 60

   Collect Data to Measure Success 60

   Issue Resolution Options 60

   Assess the Viable Solutions for Cost/Benefit, Time to Implement, and Likelihood of Success 61

   Choose the Most Value-Added Option to Implement 61

   Engage Your Employees to Help Implement the Solution—Obtain Their Support 62

   Solicit Total Experience Stakeholder Feedback 63

   Adjust the Process and Reassess Your Success against Your Goals. . . 63

   Choose the Next Pain Point and Attack It in the Same Manner...... 64

   Chapter 7 Summary 65

8 Key Performance Indicators Needed for Success 67

   Chapter 8 Summary 71

9 Total Experience Staffing...................... 73

   If Timely Decisions are Important for Growth, How Can You Make the Best Staffing Decision? 74

   What If Your Business Is Not Growing, but Shrinking?........... 79

   How Can You Best Influence Total Experience through Staffing? 80

   Chapter 9 Summary 80

10 Total Experience Mistakes to Avoid 81

   Which Business Segment Tends to Avoid Total Experience Mistakes? 83

   What about the Management of KPIs? 86

   What Happens If You Rely on Stale Total Experience Data?........ 87

   How Do Mergers and Acquisitions Affect Total Experience? 88

   Understand the Business Problem Before Selecting a Solution 90

   Chapter 10 Summary 91

11 Artificial Intelligence and Total Experience 93
What Is the Historical Importance of the AI Gamesmanship? 94
What Are the Next Steps in This Evolution? 94
Why Is This Important to Business Leaders? 96
Need a Historical Example? 97
Should Business Owners Have Cybersecurity Concerns with
AI Engines? 97
Can You Leverage AI for Data Comparisons or Reporting and Still
Maintain Cybersecurity? 98
Do Cloud-Based Total Experience Applications Leverage AI
Processing? 99
Do Business Owners Have Any Legal Relief with Respect to AI Use
and Data Privacy? 99
Chapter 11 Summary 100

12 The Future of Total Experience 103
What Are Some Examples of This Evolving Landscape? 103
What Near-Term Technologies, Other than AI, Will Influence Total
Experience? 108
Will Blockchain Expertise Be a Requirement for Business Owners
and Leaders? 109
What about Mixed Reality? 109
Fun Stuff, but How Does It Apply to a Business Setting? 110
Any Ancillary Concerns Related to Total Experience and the
Evolving Technology? 111
A Wish and a Prayer 112
Does Total Experience Really Matter? 113
Chapter 12 Summary 113

Glossary 115
Bibliography 125
Index 127
Acknowledgments 137
Let's Connect! 139
About the Author 141

INTRODUCTION

Why It's Critical to Embrace Total Experience Now

Companies like Bed Bath & Beyond, Rite Aid, Tuesday Morning, WeWork, Yellow Corporation and numerous small businesses across the United States all have something in common in 2023: *bankruptcy*. Few, unlike FTX, were involved in any corruption or illegal activities. Rather, their business ecosystem changed, they lost sight of their customer needs, or they disenfranchised their employees. Assuming these failures were not all predestined, then the management teams must have missed opportunities to alter their business trajectory. I propose that the business failures were not due to lack of skill, resources, or purposeful incompetence. Rather, they were due to the lack of a winning business methodology. Such a model for success exists but is not well known.

At the beginning of 2023, an internet search for media describing the business methodology of "Total Experience" met with limited success. While many books exist covering the individual elements of this business approach, the list of aggregated content for Total Experience was extremely limited. This is not entirely surprising. The philosophy was introduced in the forward-looking article: *Gartner Top Strategic Technology Trends for 2021* by Kasey Panetta (October 19, 2020). Gartner Inc. is a world-renowned research corporation that focuses on business concepts and technology offerings that help to address common issues.

Additionally, where authors of this article and other books explored segmented concepts of Customer Experience or Operational Excellence, very few dive into the software application components (in plain English) to answer the questions like:

- What really is this concept?
- Why is it important to my business?
- What software or solutions are available to support this concept?

This book has been crafted to fill that void.

The blessing/curse of having a career spanning three decades is that you have observed a lot of business models, leadership styles, and cultures. In writing this text, I have leveraged many of those business experiences to highlight how any organization can (and should) apply the Total Experience methodology. Further, I leaned into several real-world examples of why this concept is important to businesses from start-up to enterprise maturity.

Continuing with this thought, I have been blessed to work in the shadows of some industry icons, several are noted in this text. Many of these leaders applied Total Experience principles long before it became fashionable. Unfortunately, not all those experiences and observations were success stories. Tenures at Gateway Inc. and U.S. Leather provided valuable insights into how *not* to approach Total Experience. Conversely, involvement with eBay,

Home Instead, IDEX Corporation, PayPal, and other successful entities highlighted the tangible benefits of adopting the management principles of Total Experience.

Lastly, unusual content for this book was provided by the privilege of serving on the customer advisory boards for two Customer Experience industry giants: Genesys and AT&T. Those appointments offered insight into new products and processes long before they were publicly available. The advisory boards reviewed multiple products, many of which found their way into the Customer Experience market. Interestingly, a few were halted or significantly modified based on board feedback.

What Will You Get Out of This Book?

After reading this book, you will have a firm understanding of the Total Experience methodology. Further, you will be able to comfortably and effectively apply the concept in a real-life business setting. You will be able to speak with confidence about the principals and hold your own with vendors wanting to sell you Total Experience products.

Additionally, I have woven anecdotes and lessons from my real-world career involving learnings from leading tens of different teams in many different business settings. While this book is not intended to be a guide to leadership and management, you may find a few nuggets of wisdom and inspiration to aid your leadership journey.

Lastly, this book addresses the issues associated with growth—healthy growth resulting from mastering the balance between maximizing Total Experience and minimizing cost inputs. This is not a text on how to downsize your business.

What Is the Structure of This Book?

Chapter 1 describes the concept of Total Experience, its components and why it is important to businesses going forward. This chapter introduces you to a new business trying to scale its operations and how the Total Experience methodology and technologies can provide value. You will see that example referenced multiple times in future chapters.

Chapter 2 reviews the common technologies associated with Total Experience and how (or when) a business should consider their adoption. Additionally, the discussion delves into the automation elements of Total Experience, emphasizing the importance of technology to shape overall business interactions, including contact channels, platforms, applications, workflow tools, and survey instruments. The narrative concludes by noting the industry-specific priorities of Total Experience and highlights the role of Contact Center as a Service to efficiently manage inbound voice traffic for businesses of all sizes, even very small businesses.

Chapter 3 reviews the Customer Experience industry, the software elements that support it, and why most businesses need some portion of the functionality. The chapter also reviews an approach to adopting varying degrees of Customer Experience sophistication. (It's helpful to know that many software vendors use the acronym CX for Customer Experience.)

Chapter 4 dives into the concept of Employee Experience and Workforce Management. It provides examples of why it is critical for organizations of all sizes to consider solutions. It offers some strategies to address your scheduling and staff management needs, whether your organization is quite small or quite large.

Chapter 5 reviews the attributes of Multi-User Experience and highlights a common management tool called a Request for Proposal to compare various vendors' technology solutions.

Chapter 6 reviews the concept of Operational Excellence as a critical component of Total Experience. The chapter provides insight into the need to process data, so your business can access and improve Operational Excellence.

Chapter 7 offers a series of steps to begin a Total Experience enhancement program. This chapter reviews each step in detail for practical application within a business setting.

Chapter 8 covers the use of Key Performance Indicators (KPIs) and how they augment maximizing Total Experience. The narrative concludes by noting the industry-specific priorities of Total Experience.

Chapter 9 reviews the staffing and leadership decisions that promote Total Experience. This chapter analyzes examples of best-in-class management styles versus those that diminish Total Experience.

Chapter 10 explores a few of the pitfalls to avoid when implementing programs to maximize Total Experience in your organization. The narrative discusses the risk of overemphasizing specific business enhancement programs. The importance of avoiding being "penny-wise but pound-foolish" is illustrated through a case of a business owner who was reluctant to increase labor rates, leading to high attrition costs and other related issues.

Chapter 11 reviews the current landscape of artificial intelligence (AI) and how it can enhance Total Experience. The analysis provides examples of how businesses of all sizes are already enjoying the competitive benefits of AI.

Chapter 12 looks at the future of Total Experience and how innovative technologies will influence businesses during the remainder of this decade and into the next. Two emerging technologies are explored: blockchain and mixed reality. It summarizes the need for Total Experience and makes a plea to embrace business ethics when leaning into the adoption of Total Experience methodology.

Who Is the Intended Audience for This Book?

The content of this book is applicable to businesses of all sizes. The highlighted methodology examples are represented by small, midsize, and enterprise organizations. Further, the glossary of terms and index will facilitate easy follow-up and reference. As such, I recommend this book to:

- Small and midsize business owners and leaders
- Customer service managers
- Human resource managers
- Information technology managers
- Project or program managers
- Procurement managers
- Financial managers
- Graduate and undergraduate students

In summary, anyone who has an interest in either managing or enhancing the operations of a business will find value in this book.

CHAPTER 1

What Is Total Experience?

Maggie Chapel spent her entire life living in Nebraska and graduated from a local nursing college. Like many of her college classmates, she had dreams of graduating, passing her boards, and establishing a nursing career at one of Omaha's hospitals or clinics.

Maggie soon learned a life lesson: achieving your dreams is not always as sweet as the journey. She found herself hating her job—largely because her hospital floor was habitually understaffed. The hospital administration assured everyone: "We are recruiting new teammates as aggressively as possible!" They attributed their lack of success to a tight labor market.

While out with some friends who were also nurses, she asked: "How much of an hourly increase would it take to get you to consider changing jobs?" Most responded that a $5 per hour increase would be enough to entice a job change.

Interesting. If her sample of nurses was representative of others in the area, then a means of generating movement in the nurse workforce clearly did exist. So, why couldn't her hospital attract more nurses just by paying a little more? The reality was they could. However, the local hospitals and clinics did not want to get into a bidding war with each other. While it was perfectly acceptable to poach from organizations in other cities, counties, or even from out-of-state, it was not acceptable to poach within the Omaha metro area.

Maggie's next big "ah-ha" came from her aunt who lived in a small town in central Nebraska. As a hospital administrator, Aunt Betty had influenced her decision to choose the nursing profession. Maggie launched the conversation with: "Hey Aunt B! How are you?"

Cheerful as always, Betty replied: "I'm doing well. However, I'm wearing out my fingers dialing for nurses. We are struggling to find fill-ins for the hospital. It has gotten to the point that our folks cannot go on a week's vacation."

"Oh my, that's awful."

"Say, how much vacation do you get? Would you be interested in making some money on the side? I am desperate. We could pay $100 an hour for a forty-hour commitment."

"Whoa! That is three times my hourly rate! Do you think other hospitals would pay a premium like that?"

"Absolutely! I have several friends who work in other hospitals and clinics fighting this same battle."

Maggie's alarm bell rang. She could make a better living as a traveling nurse, plus this might be a business opportunity. She could connect other nurses with these part-time assignments. And just imagine, if her business idea really took off, she could hire a couple of nurses full time and create a regular rotation schedule!

Fast-forward six months. Maggie's business idea has soared. She has added six full-time and ten part-time nurses to her company to perform "traveling nurse coverage." She has two office assistants answering the phone and helping her coordinate customer activity. Still, she is drowning in contacts. Not only are existing clients calling, but she has numerous new institutions calling as well. They all want access to her in-demand services. Further adding to the chaos, the amount of activity required for nurse scheduling, hiring, and general office management is overwhelming. Maggie is working more than eighty hours a week trying to get ahead of it all. Her office staff is getting burned out and are threatening to quit. Her customers and her nurses are complaining about their struggle to coordinate service activities. Maggie's "Total Experience" is failing. She needs to get it right, quickly. Otherwise, a competitor will appear, solve the issue, and stifle her business.

This book is not about Maggie and her struggles. Rather, Maggie represents a wide variety of business owners. Maggie's business condition will help us explore Total Experience concepts for virtually any business in any industry, whether it is small, midsized, or an enterprise.

In 2020, Gartner Inc. was the first to seriously promote the Total Experience methodology. The company is a leading research and advisory firm providing insight, advice, and tools to businesses across various industries. For those new to the corporate software industry, companies like Gartner and Forrester Research, Inc. provide useful value comparisons between various software platforms. Likewise, they annually interview thousands of industry leaders to assess market needs facing midsize to large corporations. Gartner is credited with being the first research organization to qualify the blended impact of Customer Experience, Employee Experience, Multi-User Experience, and Operational Excellence into one methodology: Total Experience.

Who Else Should You Be Concerned about Besides Customers and Employees?

In looking at Maggie's business, we have briefly discussed the challenges she faces with her customers (Customer Experience) and her employees (Employee Experience). Who else is there? The Multi-User Experience answer comes down to the type of business you are in and the third parties who interact with you. For example, Maggie's nursing business is regulated by the State of Nebraska. This government agency would be considered a multi-user for her business. Similarly, her liability and malpractice insurance company may require specialized reporting. Further, she hired an accounting agency to help her with bookkeeping and tax reporting preparation. All these examples are third-party vendors or organizations that are particularly important to the success of Maggie's business and represent multi-user entities. Depending on your business model, the list of third parties can get extremely long.

Multi-User Experience does not receive a great deal of attention in most business books. However, most business owners will tell you that their external business relationships with bankers, referral groups, landlords, and vendors are equally important to that of their employees. Few businesses will survive and grow without strong external relationships. This especially applies to your vendor relationships. If your computer or telephone systems are routinely failing or they are not able to scale with your business growth, not having a good relationship with these vendors can be catastrophic. The same applies to product suppliers of retail businesses. Empty shelves make no revenue.

The last element of Total Experience is interesting. The Operational Excellence concept means you are striving for continuous improvement in all areas of your business and its processes. Some of these improvements may benefit current customers and/or employees, but some may not. It is possible that they may only foster future business opportunities. An example of a strategic improvement might be tied to employee training. It is not unusual for companies to train staff on products long before they are offered for sale. Similarly, information technology leaders are always looking at system scalability and disaster recovery capabilities. Expenditures in those areas will enhance Total Experience by aiding the continuation of operations during emergencies or times of disasters. As such, the redundant systems do not aid current customer or employee needs now, but they are vital during periods of primary platform failures.

As a singular statement, Total Experience is: *Anything a customer deems important to your business, anything that facilitates your employees' ability to provide a valuable product or service, anything that might influence a third-party's opinion of your business, and anything that will enable you to continuously improve and grow your business.*

So, As You Improve Your Total Experience, You Should See an Immediate Business Benefit, Right?

Generally, yes. However, with everything in life, there is a caveat. You must find the right cost input and benefit balance. Often, this is where the introduction of technology can help reduce costs and improve overall Total Experience. Likewise, adding the right staff at the right time will also benefit Total Experience. However, those new employees will take a little while to become grounded in the business processes and become strong contributors. When you get the value proposition balance right, you should experience positive influence in many areas of your business. Including:

- Customer engagement
- Brand loyalty
- Employee retention
- Higher quality products and/or services
- Reduced cost of product delivery and/or operational support

Why Is the Total Experience Business Concept New?

It is not entirely new. However, huge accolades to the Gartner Advisors for highlighting the aggregated concept. Leaders like John McCabe (Wachovia, Wells Fargo, and PayPal) have been embracing elements of the Total Experience principles for years. In 2010, upon joining PayPal, McCabe described the customer service model akin to a "three-legged stool." The legs were identified as the Customer Experience, the Employee Experience, and business performance. McCabe's focus on his interpretation of the customer service model generated excellent results. PayPal's profitability and size grew multiple times between 2010 and 2017—while customer satisfaction more than doubled.

A more recent example of a company embracing Total Experience is Tesla. Elon Musk (a founder of PayPal, SpaceX, Tesla, and OpenAI) demonstrated a relentless focus on Total Experience. In 2012, he introduced Tesla as a mainstream automotive supplier via the Model S, an electric vehicle (EV). Notably, the concept of EVs goes back to the early 1900s. However, prior to Tesla the manufacture of electricity-driven automobiles was limited.

Before introduction, Musk realized the method of purchase would need to change to spur adoption by the buying public. He made purchasing a Tesla as easy as purchasing an Apple iPhone.

His approach: configure the car online via the Tesla website and deliver it to the customer's location. To better control the quality of the purchase and ownership experience, he eliminated the dealership model entirely. To obtain service for the vehicles Tesla owners go to a company-owned service center or coordinate with on-call, local technicians.

Knowing the electric vehicle would be useless without a national recharging strategy, he developed numerous charging stations across the United States. Further, his team developed and distributed home charging equipment before the first mass-production vehicle was officially available. Musk demonstrated a passion for Tesla's Total Experience that is rare in tenured enterprises, let alone a start-up entity.

Is the Total Experience Concept Only Applicable to For-Profit Businesses?

No. It's vitally important for both profit-based companies and nonprofits to maximize their organization's Total Experience. When applying the concept to nonprofit entities, you would embrace donors in the same manner that a for-profit entity would embrace their customers. The goal is minimizing donors' frustration associated with interacting with your organization and maximizing their feeling toward your mission (Customer Experience). The effort to maximize your employee or volunteer workers' experience remains the same for both for-profit and nonprofit organizations (Employee Experience). The rest of the Total Experience concept, addressing Multi-User Experience and Operational Excellence, aligns equally well.

In summary, every aspect of a company or organization is part of its Total Experience. An old commercial for Prego spaghetti sauce had a tagline: "It's in there." The same applies to

Total Experience: "If it affects your customers, employees, third-party relationships, or your operations, it's in Total Experience."

CHAPTER 1 SUMMARY

- Total Experience represents:
  - Customer Experience
  - Employee Experience
  - Multi-User Experience
  - Operational Excellence
- Total Experience addresses all aspects of your business in a related fashion.
- Business benefits from embracing a Total Experience effort span the opinions and actions of customers, employees, third parties, and associated entities.

CHAPTER 2

Which Technologies Are Part of Total Experience?

In 2008, not all corporate meltdowns were tied to the financial service industry. Many retail companies were experiencing a dramatic shift in the expectations of their customers. eBay was no exception. Since the launch of Pierre Omidyar's eBay in 1995, the company had been addressing buyer and merchant customer service issues via email. If you were a high-value buyer (or merchant), your email received priority, with a possible two-hour response time. Unfortunately, most customers had to endure a minimum of a twenty-four-hour wait for a reply. Others, from low-value markets, might have to wait up to three days! To compound matters, resolving issues often required multiple exchanges to find an acceptable solution. This added days or weeks to the Customer Experience with eBay's customer support.

In the years ahead of 2007, hundreds of thousands of new customers were joining the eBay marketplace monthly. Revenue growth was so steady that eBay was considered a bellwether stock. However, that was now changing. At the annual executive strategic offsite the alarm bells were ringing. Company revenue for 2007 was projected to be $7.7 billion. The forecast for 2008 was shaping up to be flat compared to 2007. Moreover, 2009 was projected to be down by 3 percent. Something had to be done!

eBay reacted by promoting John Donahoe to president and CEO at the beginning of 2008 following Meg Whitman's resignation. One of his first actions was to review eBay's Customer Experience from the lens of a casual buyer and from a merchant on the platform. At the time, eBay employees knew Donahoe was unimpressed with the marketplace website's search capability. He was equally disappointed by the customer support offered to eBay merchants and buyers. The fact that 70 percent of the customer inquiries were satisfied by slow-response email was deemed unacceptable. He noted that eBay gave premier customers a phone number or a web chat option. Those customers, when polled, offered better satisfaction scores for the company. However, even those contact channels often experienced clunky transfers and call drops. Overall, Donahoe found that eBay was driving away tenured customers faster than it could add new clients to the platform.

In early 2008, the company initiated a full-court press to shift customers from email to the telephone contact method. This meant adopting the customer service model leveraged by PayPal, a wholly owned subsidiary of eBay. PayPal's approach directed 70 percent of its daily contacts to the voice channel, with merchants and buyers reaching out to customer service agents by telephone. Not surprisingly, the financial transaction subsidiary had better Net Promoter Scores for its customer base than its parent company, eBay. (Companies use the Net Promoter Score polling metric to measure customer loyalty to a product or service.) Further, PayPal had higher retention rates for its customer service teammates.

Say What? PayPal Had Better Employee Retention Rates?

Yes. The technology tool sets made inbound voice inquiries easier to support. The speed of resolution provided better post-contact customer feedback, making PayPal teammates more apt to stick around. They received better customer feedback scores and found the technology experience to be easier.

In early 2008, eBay launched the reengineering of the customer service model under the internal code name Project Turnaround. Within four months, the company's contact distribution underwent a dramatic transformation. Customer service agents shifted to processing 70 percent of their contacts via the voice channel and 12 percent via web chat. They processed the remainder (principally international contacts) via email.

Immediately, eBay experienced an improved Net Promoter Score. The loss of tenured customers slowed, leading to enhanced revenue performance. eBay employees also felt the positive changes—as evidenced by improved retention rates. By the end of 2008, the enhancements to the eBay website search functionality and the focus on the Total Experience components had improved company performance. The following year, several magazine articles featured John Donahoe's achievements. They all highlighted the remarkable positive movement of eBay.

So, What Did Donahoe Focus On?

He reaffirmed that positive Customer Experience has a direct correlation with corporate growth. Likewise, while not explicitly called out, he appreciated the fact that employee mindsets had an enormous impact on overall business performance. Crappy customer service tools, irate customers, and a lack of agent control generate unhappy customer service agents. Moreover, Donahoe knew companies spend substantial time and money recruiting new service staff to replace disgruntled teammates who leave for greener employment pastures.

Donahoe also recognized that the company's technology strategy could have a tremendously positive influence on the behavior of customers, employees, and associated third parties. He was a Total Experience leadership champion long before it was fashionable.

What Are the Automation Elements of Total Experience?

We have already established that anything the customer, the employee, or an affiliated entity touches affects the overall business experience. This includes:

- The contact channels used to purchase and review products or services
- The platform used to record the sale
- The applications used to provide an excellent quality product or service
- The workflow applications used to track customer issues
- The data and reporting platforms for customer, internal business, and external agency needs
- Customer and employee survey tools

It is worth noting that even something as obscure as Identity and Access Management software can be a key component of Total Experience. Companies leverage this software to grant customers security permissions on their websites. Surprisingly, for many years, the number-one contact reason for customer contacts at eBay and PayPal was assistance with account password resets. At eBay, this demand for support typically surged around October and November each year when Christmas shoppers returned to the marketplace for their "annual present." Those shoppers required help restoring their access to the website. Eventually, both eBay and PayPal leveraged modern technology to implement self-service Identity and Access Management capabilities, enabling customers to reset their own passwords.

Are the Elements of Total Experience Applications the Same Across All Businesses?

No. Total Experience process priorities hinge on your industry or business model. For a concrete delivery business, the focus might be on manufacturing and delivery coordination applications. In contrast, a dog daycare business may need to concentrate on a platform to record customer information, details about the dogs, and the service attributes customers routinely want performed.

Most businesses share some common functions. For example, every business today must have an email platform and an associated email account. Another necessity for every serious business is an invoicing/bookkeeping platform, following a rule offered by business consultant and author Peter Drucker: "If you can't invoice, and can't effectively manage collections, you won't be in business very long."

Returning to Maggie and her office turmoil, we can assume she has the invoicing and bookkeeping activities handled. However, she needs help triaging inbound contacts, both from her clients and her employed nurses. In the 1970s through early 1990s, this technology automation area was dedicated to supporting large corporate enterprises. Companies like Avaya, Aspect, Cisco, and others were introducing new Private Branch Exchange platforms that allowed companies to triage their inbound telephony contacts and apply Interactive Voice Response systems. The combination of software capabilities helped manage the telephony volume. The cost of this technology restricted its usage to *Fortune* 500 companies. However, in the 2000s, many new market entrants brought the same capabilities to smaller business customers for a fraction of the cost. This influx of cloud-based software vendors spawned a new industry segment: Contact Center as a Service.

For today's business owner, having the ability to easily program an Interactive Voice Response system to support inbound customer calls is especially helpful. You can configure these platforms to provide a certain Customer Experience during normal business hours and a different one outside of normal operations. You can keep it simple with a recorded welcome message and the rudimentary selection for a service option. For example, we have all experienced: "Press 1 for sales, 2 for service, or 3 to leave a voicemail recording."

The more sophisticated systems allow callers to ask their question in a conversational manner, known as *natural language*. The computer application interprets the customer's question, and the call is dispositioned to a customer service agent. Even more advanced software platforms will authenticate callers to their user accounts and allow for self-service activity. Many banks use this capability to address customer account balance inquiries and other common functions like password resets. This is all completed without having to engage a human customer service agent. Most of the Contact Center as a Service software vendors have this capability available on their platforms, offering businesses of all sizes tremendous flexibility to manage inbound voice traffic.

In summary, well-designed and well-implemented Total Experience technology platforms will have your customers, employees, and affiliates applauding your business. Conversely, technology adding to frustration may drive customers and employees away.

CHAPTER 2 SUMMARY

- Total Experience business technology encompasses any digitalization of processes that affect customers, employees, the business activity, or the support of a third party.
- Any business, large or small, must be mindful of Total Experience for continued growth, sustainability, and cost management.
- The Total Experience focal points and associated solutions vary based on the specific business needs of each organization.

CHAPTER 3

Customer Experience

The Customer Experience industry has grown extensively over the last forty years. The technology it encompasses varies by vendor. However, all vendors in the space can handle the management of customer contacts via various sources, referred to as *contact channels*. Those contact channels include inbound or outbound telephone calls, email, web chat, and even casework assignments. The last channel, casework assignments, may or may not be initiated by a customer contact. That is, it could be any event that needs to be reviewed by an automated workflow or a human processor. An example of a case for a financial institution could be an alert that was triggered by monitoring all financial transactions for specific criteria that might indicate a fraud event. Upon identifying a case, the Customer Experience platform may be programmed to transfer the case to a risk agent for action or initiate some automated action. The more advanced platforms also manage outbound call campaigns and text messaging between customers and business staff.

Additionally, the more capable Customer Experience applications have robust reporting functionality. The software offers either detailed published reports or dashboards with drill-down capability. Managers use this capability to assess the aggregate performance of a contact center's teammates, identify training needs, and to spot customer trends. Even the most basic applications can support operational performance indicators like:

- How many calls are on hold or in progress?
- How many emails have been processed today versus yesterday?
- How long does it take to respond to the chats?
- How many chats are active per signed-in customer service agent?

Some vendors have extensive modularization of their product offerings, which allows customer service agents to upsell features. Furthermore, many platforms charge to integrate their products with other applications like financial accounting systems. As you might imagine, some of those out-of-the-box integrations are sold at an enormously high price.

In the 1980s, companies like Aspect and Avaya were open to offering "concurrent seat" licensing structures for their Customer Experience technology solutions. With that model, you did not purchase a license for every named user. Instead, you only bought enough licenses to support the maximum number of "concurrent agents" logged into the system during peak periods of activity. Unfortunately, most software companies have moved away from that model, finding it more lucrative to sell named user licenses. Under a named user structure, you need an individual license for every agent who will be taking a call, chatting, or responding to an email. Or, more generally, anyone needing to be logged into the Customer Experience platform requires a license.

Today's software vendors have been transitioning their product offerings to the cloud via Contact Center as a Service platforms.

With cloud offerings, you can access the same features available from the software that you would otherwise have to install on servers within your own office environment. The huge benefit is not having to worry about security patching the hardware, increasing the scalability of the infrastructure, or having to perform routine application upgrades. Rather, these tasks are performed for you as part of the service offering.

The new Customer Experience vendors have driven down the licensing costs over the last thirty years. Since 2010, cloud application licenses have been offered between $75 and $170 per seat per month. That sounds expensive until you consider the cost of adding customer service labor. However, Customer Experience licensing just provides access to the software features allowing for efficient triage of inbound or outbound contacts. On top of that, you still need to pay your local telecom company for telephone numbers, text messaging, and service usage. The same typically applies to email providers and email IDs.

Today's "500-pound gorilla" in the Contact Center as a Service industry is Amazon. The twist that they bring to the game is a pricing model that is usage based. Their subscription costs also include the telecommunication expense you would normally incur with AT&T or one of the other carriers for inbound voice calls. If you want to enable Customer Experience management for your inbound voice traffic via Amazon Connect, great! You can do it for less than $0.02 per minute. If you want to leverage Amazon Connect's chat service on your website, awesome! You can enable it for $0.004 per message. The services allow you to blend channel offerings and add other self-service features like SMS texting. This cafeteria pricing model is consistent for all Amazon Connect service offerings (both inbound and outbound traffic). For many businesses, this pay-as-you-go pricing model is extremely attractive. It allows a business owner or leader to align and forecast contact center management costs with business volume.

Many small and midsize business owners have chosen to capitalize on another Customer Experience opportunity: outsourced customer management services. This represents hiring a professional answering service to take your calls and capture customer information. Notably, this answering service concept is not new. It has been around since the 1920s.

What is different is that the new vendors leverage the capabilities of the Customer Experience software platforms. This makes their operations significantly more efficient at providing the services for many hundreds or thousands of small businesses. A well-known player in the customer contact service space is Ruby. This vendor offers live-agent service plans based on the volume of receptionist minutes utilized each month, ranging from $3 to $4 per minute. Ruby's service offerings can answer customers' questions, help them schedule appointments, and collect new customer information for follow-up by the business staff.

Other vendors in this space include:

- AnswerConnect
- Answering Service Care, LLC
- Smith.AI, Inc.
- SureCall Experts

As you may be able to tell by some of the vendor names, many service providers are focused on live, human reception services augmented by technology. Some vendors provide artificial intelligence-based service offerings augmented by human support. The type of service you may require is a question of the best fit for your specific business and cost structure. As you can imagine, the more "white-glove," human-based service offering you lean toward, the higher the cost per contact. If your average client inquiry call takes more than ten minutes to complete, this reception-based service can get expensive, quickly.

For Customer Experience applications to function effectively, they typically need to be connected to a Customer Relationship Management system. The two applications work in concert to help prioritize the contact response via customer phone number, email address, IP address, or some unique identifier. The Customer Relationship Management platform concept has existed since the 1950s (the first was an electronic Rolodex). However, the explosion of cloud software offerings also started in the late 1990s (in conjunction with Customer Experience application development) and accelerated in the mid-2000s. The products were originally targeted to support midsize to large enterprises.

This left start-up and small business owners to leverage spreadsheet applications to track basic customer information versus implementing a formal Customer Relationship Management platform. The tracked content usually included customer names, addresses, and contact information. The spreadsheet approach was cheap to implement and helpful in the early days. However, these ad hoc Customer Relationship Management systems soon became clunky. Owners quickly found the need to have a system integrated with a point-of-sale system, a billing/bookkeeping platform, and/or a Customer Experience application. The customer-management concept really took off with Salesforce in 1999. This software is attributed to being the first large-scale, cloud-based application.

Upon introduction, the Salesforce selling point was: "It is so easy to enable that you don't need the help of the IT department." Their claim was certainly true for the initial configuration. However, as soon as you wanted to integrate their software with another application, the IT department was quickly identified as a critical collaborator.

Today, oodles of companies compete in the Customer Relationship Management space. A few of the more aggressive vendors include:

- HubSpot, Inc.
- Microsoft Dynamics 365 Customer Service
- Monday.com
- Oracle Siebel
- ZenDesk

Pricing models still revolve around a "per agent seat" structure. However, for as little as $14 per month, you can get basic software capability to aid in tracking customer attributes and contact information.

Over the last ten years, three new industry shifts have been occurring. First, the Customer Relationship Management applications have been adding capabilities that bleed over into the Customer Experience space (and vice versa).

Second, niche players have been gaining traction. For example, if you run a nonprofit organization and need a donor-tracking system, you can use a generic customer relationship application (like Salesforce). However, you may find better value by targeting nonprofit donation management applications like Bloomerang, DonorPerfect, Givebutter, or others. The same applies to other industries like automobile sales or real estate businesses.

The third major shift is that Customer Relationship Management vendors have been entering the digital payments space. This is happening with very tight integrations with platforms like Adyen, Apple Pay, Block, and PayPal.

An important area of Customer Experience applications worth noting is the ongoing push to provide customer and employee self-service opportunities. We mentioned the Tesla automotive buying experience whereby customers can configure the options of their vehicle and choose the location/delivery method that best meets their needs. This customer self-service functionality is being heavily embraced by all the Customer Experience software manufacturers. For customers dialing in on the telephone, this is encompassed using an Interactive Voice Response system.

For those clients leveraging web chat to contact customer service, often a chatbot is the first interaction point. It leverages a list of frequently asked questions to try to satisfy the customer's inquiry without having to engage a human. At the end of the last decade, the cost differential between having an auto-attendant handle the contacts versus a human was quite extreme. In the telecom space, where companies typically had a twelve-minute average handle time for a human, the cost per contact for a software-supported contact was often ten times cheaper than that of a live agent processing a call. This savings difference considers licensing, engineering support, and equipment cost for the software platform versus the live agent's salary and benefits. Unfortunately, the customer satisfaction survey results for those events were often negative. Depending on age, demographic, and other considerations, some customers just want to speak to a live agent.

This customer satisfaction survey space is worth a brief sidebar discussion. This is a software and cloud application industry unto itself. There is a science to asking the right questions, in the right order, and with the right verbiage to obtain actionable sentiment from your clients and employees. The enterprise players in this space have been Alchemer, Medallia, and Qaultrics. They dominated the landscape for several years. Now, many new cloud-based companies are severely disrupting the market. This is especially the case with applications geared to small and midsize businesses. These cloud-based customer satisfaction survey applications include:

- CX Index
- Forsta (previously Confirmit)
- SurveyMonkey
- Zonka Feedback

Do All Customer Experience Applications Have the Same Features?

No. The capabilities of the applications vary widely. For example, some software products offer self-help features that speed the training of employees on specific functions. This may include a question bar that allows the employee to type a conversational question while interacting with the customer. The Customer Experience application will query a knowledge base of information pertinent to the business policy or process questioned and offer application choices to satisfy the question. Other applications have extensive how-to video libraries that customer service employees can reference from within the application based on specific topic areas.

As you may be able to gather from above, the complexity of the client-support application topology can become messy. This is why many vendors are trying to incorporate more of the ancillary capabilities into their core suite of products. Figure 3.1 provides a basic depiction of the Total Experience activities the Customer Experience vendors are trying to support.

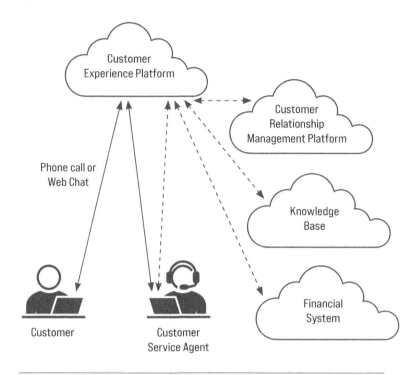

FIGURE 3.1 Depiction of the Total Experience activities the Customer Experience vendors are trying to support

The solid lines represent the real-time communication between the customer and the customer service teammate. The dashed lines represent data transfer used to validate the identity of the customer and leveraged to support issue resolution. This may include transactional information related to the customer's purchase or payment activity.

How Can a Business Owner Achieve Good Technology Value?

You need to understand your specific use cases and business needs. It is critical to select your software products carefully to avoid purchasing duplicated features or overlapping capabilities. Lastly, please recognize that your services and tool sets will change over time. You may want to start with a vendor-managed service offering. At some point, your business will grow to a point where it will be more cost effective to add the technology and hire the talent needed to manage it internally. Even after selecting a software product to manage in-house, you need to periodically review it every two or three years to ensure it is still meeting your business needs in the most cost-effective manner. Many older organizations have acquired applications that are so negatively entrenched in their processes that they represent "technology debt" for the organizations.

CHAPTER 3 SUMMARY

- Customer Experience platforms vary widely in cost, capabilities, and integration abilities.
- Customer support applications require the capabilities of Customer Relationship Management platforms to maintain client information.
- The modern Customer Experience platforms offer customer contacts via telecommunications, chat, email, or website interaction.
- Cost-effective Total Experience workflows require selecting software products that do not overlap application features.
- Business needs change over time, which requires you to periodically review existing applications for current business value.

CHAPTER 4

Employee Experience

Imagine an operations center that consists of 600,000 square feet of office space. To put that into context, this facility houses fourteen acres of cubicles, control center operations, data center equipment, and conference rooms. That is the size of the AT&T Operations Center in Dallas, Texas. Before the COVID-19 pandemic, it housed over 4,000 engineers and customer service staff.

The AT&T center, responsible for monitoring the performance of over one million miles of fiber optic cable, more than one hundred data centers, and thousands of cell towers, also houses a Security Operations Center. This segment of the facility resembled the bridge of the starship *Enterprise* from *Star Trek*. It had wall-to-wall monitors staged to share an alert if a concern arose for the AT&T system or its millions of customers. The Security Operations Center monitors cybersecurity threats or breach activity. During a tour, AT&T leadership highlighted cyber hacking attempts launched by a mixture of governments, criminal organizations, and individuals. Everyone on the tour was amazed at AT&T's ability to confidently identify the attackers. Security personnel determined the hackers' identities by the method of attack, internet protocol (IP) addresses, the size of the attack, and several other event signatures. It was truly impressive.

Now, this was just one of the AT&T service centers within the corporate structure. This mega-corporation has several service centers across the U.S. and in international locations. Aggregating all facilities, the company employs tens of thousands of customer service agents who process hundreds of millions of customer contacts each year.

How Does AT&T Manage All the Diverse Needs of Its Employees?

Managing employees' needs begins with understanding and documenting them. While payroll and benefits are clear requirements, many companies often overlook other aspects like career aspirations, training objectives, and performance reviews. Thankfully for the AT&T employees, the company has years of experience fostering all aspects of employee support and engagement.

In addressing payroll and benefit considerations, AT&T, like most midsize and large companies, relies on a Human Capital Management system. The system houses critical data, such as payroll information, benefit information, employment history, employee addresses, and information needed for regulatory reporting. Many platforms also include organization charts and access to training content. Leading software platforms in this field include Ceridian, Oracle, SAP, UKG, Workday, and various other smaller brands. Conversely, smaller companies often use a payroll-focused service provider or application to house their employee information. Those applications include ADP Payroll, OnPay, Paylocity, and a litany of other vendors that cater to small and midsize companies.

Beyond payroll and benefits, companies often conduct employee surveys, hold regular feedback sessions, identify critical business workflows, and strive to create a culture of open communication with management. Once these needs are identified, the organization may require software solutions or service providers to meet employee satisfaction targets. This can include tools for collaboration, project management, communication, and task organization. Choosing user-friendly software with intuitive interfaces to streamline workflows and enhance employee efficiency is critical to maximize Employee Experience. While large platform vendors, like Google, Microsoft, and Slack, aim to maximize employee efficiencies across businesses of all sizes, the small business market is populated with low-cost, limited-feature solutions.

Regardless of the chosen path, it is important to select a solution with robust training and support offerings. Additionally, providing information technology support ensures employees feel confident and capable of utilizing the tools to their fullest potential. Matching employee needs with appropriate solutions allows businesses like AT&T to create a work environment that maximizes productivity and fosters employee engagement.

How Does AT&T Manage the Schedules of All Customer Service Employees?

The software solution facilitating employee scheduling is called Workforce Management. Operating alongside the Customer Experience platform, Workforce Management applications track agent schedules, monitor log-in events, and provide crucial statistics, such as the average number of calls per period, average call handing time, number of emails worked per period, and other useful data.

Interestingly, Workforce Management capabilities extend beyond just employee scheduling. Enterprise-grade applications for customer contact centers capture the historical volumes for each contact channel. The platform uses this data to forecast future volumes for selected work periods (for example, minutes, hours, days, and weeks). It is this blend of agents, anticipated volumes, contact channel knowledge, and work schedules that enables companies to properly staff their contact centers. Workforce Management provides the ability to balance customer service experience while minimizing the cost of providing agents.

Unsurprisingly, the enterprise-grade applications have traditionally been developed by the top-tier Customer Experience application providers. For many years, the best Workforce Management applications were offered by Aspect, Avaya, and Cisco.

As new software vendors entered the market for small and midsize companies, they offered application suites integrated with contact management, staff scheduling, and advanced telecommunication functionality. With this influx of vendors and integrated solutions, a small business can now enjoy many of the same cost-management capabilities once restricted to large enterprises.

Do I Really Need a Workforce Management Solution?

That depends on the size of your workforce and the complexity of your Customer Experience activities. For Maggie's small office, she desperately needed a software application to handle the inbound contacts (which also dictated the need for an integrated Customer Relationship Management application). However, to maintain the schedules of her small office staff, there is a low-cost/no-cost alternative: the tried-and-true spreadsheet approach.

If she is using Microsoft Business 365 Premium for her office applications, she will also gain access to OneDrive. Leveraging a cloud-based business application suite like this allows Maggie and her office staff to easily share spreadsheet documents. The team can record staff schedules (including actual hours worked) in an Excel spreadsheet on OneDrive. Since the entire office team can access the document online, they can all view each other's scheduled work hours. Additionally, by inputting their actual work hours, Maggie can use the same document for her payroll needs.

It is worth noting that Excel offers a decent number of predeveloped business templates, making it simple to create schedule models and payroll datasets with minimal effort.

If you are not keen on Microsoft products, the same functionality is available with Google Workspace Business Plus. Google Sheets and Google Drive offer similar functionality to the capabilities outlined above. However, the predeveloped business templates in Google Sheets are more basic compared to those provided by Excel. Opting for Google means spending more time customizing the model to fit your specific requirements.

At What Point Does the "Poor Man's Spreadsheet" Option Fail and Require a Formal Application?

The administration time you spend maintaining the team's work schedule will dictate when you need to shift to a commercial Workforce Management package. The low-cost Excel or Google Workspace option is fine for simple environments where employee schedules do not change frequently. Further, if employee scheduling is not segmented by contact channel, you may be fine with the spreadsheet model. However, complexity grows quickly.

Let's look at an example. A customer service provider offered centralized web chat management for several hundred small business clients. A small team of nine centralized customer service agents managed the chat-generated contacts. This support team would respond to customer inquiries only via web chat. If a more detailed response was needed, they connected the customers with local office staff by scheduling follow-up telephone calls. The activity started as a pilot and grew to supporting approximately 400 offices. The manager for the customer service team was able to manage all the shift scheduling via Excel.

As part of a business expansion plan, the centralized service provider decided to add inbound voice support for a handful of offices in the Central Time Zone as a pilot. To support a Monday-through-Friday support schedule, the manager added four additional teammates to the web chat staff. Suddenly, the complexity of managing two contact channels via an Excel spreadsheet became much more difficult. The pilot was so successful, the company decided to offer inbound voice support to another thirty business offices across four time zones. This created the need to increase contact center staff beyond thirteen employees. The manager for the support group cried "uncle." He insisted upon leveraging a Workforce Management application integrated with the company's existing Customer Experience platform. The complexity of keeping all the teammate schedules, the hours of operation, holiday schedules, and inbound volumes became too complex for a spreadsheet model. He needed a scheduling product integrated with the customer support software to handle the tactical activity and forecast future staffing needs.

The point of this example is that Workforce Management complexity is added in many ways. Contact volume is one variable. However, team member work schedules, contact channel variation, the time zones supported, and end-customer variation all add to business complexity. In the above example, the support team could have doubled the number of web chat contacts, added a few more customer service teammates, and continued to leverage the Excel spreadsheet Workforce Management model. However, the addition of the inbound voice channel significantly increased the overall complexity of the operation.

The tipping point for when you need an integrated Customer Experience/Workforce Management solution will be extremely specific to your business activity. A general rule of thumb is to leverage the tools you currently own for as long as reasonably possible. That may be represented by a spreadsheet or a whiteboard visible to everyone. However, if you delay adopting an integrated application too long into your business growth curve, your cost structure will suffer with increased employee overtime, scheduling confusion, and disgruntled customers.

Common Considerations for Purchasing Workforce Management Software

The first question to ask is whether your Customer Experience application has a Workforce Management component available. If so, how is it priced? If your application has a module with scheduling capabilities and its feature sets can meet most of your needs, there is a strong argument for leveraging it versus purchasing a "bolt-on" alternative from another vendor. Many top Customer Experience and Workforce Management applications offer out-of-the-box integrations, which require minimal effort to integrate.

Leveraging a feature set internal to the application eliminates a significant amount of technology complexity. Likewise, if you encounter an issue, you only have one vendor to coordinate versus two. This avoids a great deal of finger-pointing between vendors and typically generates a quicker solution.

If your Customer Experience application does not offer Workforce Management functionality, but the vendor has a formal relationship with another software vendor, please give their offering consideration. This is not as ideal as having the two software functions maintained by the same engineering team. However, it is preferable to having integrations between two disparate applications and engineering organizations.

In most office and contact center environments, labor is typically one of your largest business costs. Successfully managing it can be a strategic advantage. Many businesses struggle with labor cost management, which can balloon into hidden benefits, training, and ancillary facilities costs. For enterprise-sized operations, even a few seconds of talk time on every contact can add up to millions of dollars of cost per year. Workforce Management provides the data and tools to successfully manage contact support labor.

CHAPTER 4 **SUMMARY**

- Business success is dependent upon maximizing Employee Experience with cost and business value.
- Midsize companies and large enterprises manage employee information with Human Capital Management software; small firms often just leverage their payroll solution to store the data.
- Employee needs beyond payroll and benefits require tools for collaboration, project management, communication, training, and task organization.
- Workforce Management applications track agent schedules, monitor log-in events, and forecast

future customer contact volumes and aid in staffing optimization.

- Businesses must consider their workforce size and the complexity of Customer Experience activities to determine when to adopt a formal Workforce Management solution.

CHAPTER 5

Multi-User Experience

In 2016, PayPal was enjoying its one-year anniversary as a newly separated company from eBay. The company was finally able to spread its wings and define a business strategy that was truly in its best interest. That included changing how customers could engage with the company with respect to account funding methods. During the company's history as an eBay subsidiary, the push was to have customers leverage their checking and savings accounts as the primary source of funding transactions. By performing automatic clearing house transactions—commonly known as ACH—the company curtailed its financial cost-of-doing business at the expense of customer preference. With the separation from eBay, PayPal could now offer a choice of funding instruments. Consequently, PayPal made a mad dash to generate new relationships with Discover, Mastercard, Visa, large international banks, and regional banks. Then-CEO, Dan Schulman approved and marketed this strategic decision. However, the concept and its successful execution has been largely attributed to Bill Ready (now CEO of Pinterest). The effect on the business was almost instantaneous. Before making the decision, the company was essentially trying to run a marathon with ten-pound weights tied to its shoes. However, after completing this strategy, the weights were removed, and the company was soaring!

With newfound growth and exposure, PayPal quickly recognized the critical need to support Multi-User Experience. As a subsidiary of eBay, regulatory agencies showed cursory interest in PayPal. However, by 2016, this had changed. Many new state and federal regulatory agencies began focusing attention on PayPal. In particular, the Consumer Financial Protection Bureau (CFPB), established in 2010 by the United States' passage of the Dodd-Frank Wall Street Reform and Consumer Protection Act, made its existence known to PayPal. The CFPB's mandate is to safeguard customers in the financial marketplace in a very broad manner. During 2016, the seriousness of CFPB scrutiny was evident with a $185 million penalty assigned to Wells Fargo for using widespread deceptive sales tactics.

When the CFPB turned its attention to PayPal, the leadership team took the attention *very* seriously. They quickly realized that even if the company was performing its activities correctly, legally, and fairly, there was still the potential to be fined. The company had to adequately prove its activities were lawfully compliant via reporting and data submittals to the CFPB in a timely manner. If PayPal could not provide the information fast enough, the company could be fined tens of thousands of dollars per day.

As such, PayPal's leadership allocated tremendous time and resources to meet the new multi-user's needs. They fast-tracked new reporting processes to support all aspects of the customer life cycle. This included up-to-date reporting on:

- All transactions and the fees charged
- How quickly remittances were made

- How quickly customer service calls were answered
- Any fraud events detected and how quickly they were addressed
- The customer surveys conducted and their results
- And a litany of other activity reporting on advertising, sales, and marketing practices

During 2016 and 2017, PayPal added new employee resources to help support the new regulatory agency requirements. The individuals made it a point to do more than just hand over the documentation and data; they worked with the agency to understand the data in context with the agency's policies and guidelines. Within a few months, a relationship that started as challenging began to change to one of trust. PayPal went from being a company of concern for the CFPB to being the role model for how financial transaction companies should execute on customer experiences and work with regulatory agencies.

Is Multi-User Experience Only Associated with Regulatory Concerns?

No. Any business relationship that is outside of the direct interaction with customers or employees falls into the realm of Multi-User Experience. This includes vendor relationships for all manner of products and services. The list can be quite endless, including everything from product inventory suppliers to banking and insurance relationships. Likewise, all company software applications, utilities, and facilities providers would fall into the Multi-User Experience category.

Are There Software or Service Applications Dedicated to Multi-User Experience?

Yes. However, they are generally based on specific industry needs and requirements. In the regulatory space, the agencies often have specific applications that they provide to companies to ensure data consistency in reporting. Banks and credit card networks will also define specific software applications or integration requirements. Because the category of potential stakeholders is so broad, identifying one specific application or service leveraged by all companies, large and small, is a challenge. One tool that companies have universally adopted to support the Multi-User Experience to select vendors is the Request for Proposal (RFP) document.

To RFP or Not to RFP—That Is the Question!

A Request for Proposal can be a tremendously useful document when entertaining the purchase of a new software application or service. These documents tend to force a clear understanding of your business problem. Likewise, they foster the construction of a list of requirements that the vendors need to meet to help you solve your business problem.

A rule of thumb is to structure software application RFPs with a summary description of the need, objectives, timeline, evaluation criteria, and selection process. However, it will then quickly shift into a list of questions grouped by stakeholder segments. For example, the summary might be something like this:

> "ACME Corporation, a world leader in bird studies, is seeking software application solutions that will aid in the identification and capture of roadrunner birds (*Geococcyx californianus*) for research purposes. The application will be

utilized by one coyote administrator in a value-added and safety-conscious manner. The company is seeking an application that is cloud-based and does not require a private data center to facilitate its operation. The company wants to have the application evaluation, selection, and implementation process completed by June of the coming year. The selection process will be based on completeness of solution, customer support attributes, cybersecurity, and price per user. Request for Proposal responses will need to be submitted by December 31 of this year for consideration."

Thereafter, the document will often have several questions to better evaluate differences between the vendors' products. Those questions are often grouped, so it is easy to complete the analysis. If your business is an established, midsize entity with many department stakeholders, this segmentation can help everyone review completed submittals. The topical groupings make it easier to schedule stakeholder attendees and vendor subject matter experts to review specific responses. It also aids context between the questions.

An example of those groupings might be:

- General
- Legal and compliance
- Technical
- Customer support
- Administrative features

The questions will be specific to your unique business needs. However, every analysis will include questions like:

- How do you price your product (for example, per named user license, per unit of activity, or per some other measurement)?

- Does your sales process offer a proof-of-concept trial period? If so, how long is the trial period?
- Does your application offer a reporting or dashboarding segment? If so, please provide a detailed description.
- How long does it take to implement/configure a working instance of your application?
- Do you offer free customer support 24 x 7?

The purpose of the question list is to better understand the software application, each vendor's ability to support the application, and how each vendor compares to the competitors' products. Often, companies leveraging the RFP process will aggregate all the vendor responses into a spreadsheet format to better visualize the best option. An example is depicted in Figure 5.1.

In this simple example, the shading highlights the best and/or most affordable features. You will note that the pricing method makes Vendor C the most expensive of the options based on the anticipated business demand for the service. As such, Vendor A would be the better choice—considering cost, time to implement, and a trial-period offering.

Admittedly the use of the RFP process is more prevalent with midsize or large corporations versus small businesses. These larger entities find this process helpful to fully understand the problem they are trying to solve and to generate an apples-to-apples comparison between the solutions. Likewise, it allows them to focus on the key requirements they must satisfy versus nice-to-have features. Lastly, it provides procurement teams with a wealth of data to use for contract negotiations. For example, a specific vendor may charge a fee for feature "x"; they can point out that competitors offer this feature free of charge. Often, that provides leverage for a lower license fee or for getting the vendor to add the feature at zero-cost as a concession.

Multi-User Experience | 43

Questions	Grouping	Vendor A	Vendor B	Vendor C	Clarifying Comments
Pricing	General	$175	$150	$25	
Target birds captured per month		10	10	10	
Total Cost Per Month		$175	$150	$250	
How do you price your product? (e.g., per named user license, per unit of activity, or ?)	General	per seat /mnth	per seat /mnth	Per bird captured	Vendor C caps at $200 per month
Does your sales process offer a Proof of Concept trial period?	General	Yes	No	No	
Does your application offer a reporting module?	Reporting	Yes	No	No	
How long does it take to implement/configure a working instance of your application	General	1 week	2 weeks	4 weeks	
Do you offer free customer support 24 x 7?	Support	No	Yes	Yes	

FIGURE 5.1 An example of how a company might aggregate all vendor RFP responses to compare the software applications and identify the best offering for this business need

Small business owners will benefit from performing at least one Request for Proposal. It will crystalize the questions you should be asking in all contract or purchase activities. Completing formal reviews for all your future Total Experience applications or services may be overkill.

Why Would a Formal Proposal Process Not Always Benefit a Small Entity?

It is a fact: a small business will have limited price leverage with well-established software vendors. Let's revisit Maggie's business for a moment. Unlike a large enterprise, which may garner a dedicated account representative for a specific vendor, Maggie's nursing business will be assigned someone who has fifty or sixty other clients to support. That representative may spend a few hours helping Maggie understand the product and pricing model during the sales process. However, it is doubtful that the vendor will offer a discount from the published list price. Having a Request for Proposal list of questions will certainly help Maggie review all the product/business considerations. However, it is unlikely that the account representative will take the time to complete the document. (This follows the old saying: "the juice isn't worth the squeeze.") An exception would be a representative for a new vendor with very few clients. This sales rep may embrace the proposal completion effort to gain one more client or to gain experience with completing the response process.

Should a Small Business Complete a Proposal for Service Instead of an RFP?

For small businesses that offer a service, it makes more sense to invest your time providing a proposal for a service offering to your prospective customer rather than completing a customer's Request for Proposal. If the service offered is for customer reception, yard service, software development staff, or something in that vein, then having the detailed questions and success criteria documented helps create the necessary contractual agreement.

Of note, many vendors leverage a Master Service Agreement to establish generic master service details between the two parties. Examples of those terms might include payment expectations, liability insurance needs, requirements to terminate service, and so forth. Then, for the more specific contract elements, they will utilize an addendum referred to in the Master Service Agreement as a Statement of Work.

The details represented in the proposal are often translated into the Statement of Work to define specific contractual agreements between the two parties. Those might include the number of hours of service to be included per period and the specific cost.

Does an RFP Always Increase the Odds of a Successful Total Experience Effort?

Sadly, no. Even large companies leveraging a formal procurement process can make horrible mistakes that cost tremendous time and resources. For example, in 1999 Gateway Inc. was still a major player in the personal computer industry with over 20,000 employees. Company growth was still increasing. Further, the company had just completed implementation of a Y2K-compliant enterprise software system. That platform was managing all sales, manufacturing, and customer service functions—and Y2K compliance meant it would not fail at midnight on December 31, 1999. (Much of the world was gripped by fear, because all computer systems were "supposed" to fail due to the lack of databases being able to shift from 1999 to 2000.)

During a leadership golf outing, Tom Siebel, founder of Siebel CRM, made a strong sales pitch, touting that his application could manage the Gateway sales and customer service functions multiple times better than the current platform. Gateway leadership took the bait and indicated interest. However, Siebel CRM staff would need to pass Gateway RFP scrutiny. Gateway staff would conduct a comparison analysis between the newly implemented JD Edwards World application and Siebel CRM. And just to make it interesting, the Gateway team had only two weeks to complete the analysis.

Gateway gathered its subject matter experts from sales, manufacturing, marketing, finance, and customer service. They created a multipage list of critical features based on the company's not entirely unique requirements. For example, the company purchased identical computer components from many vendors. It needed to identify which vendor's component was used in each personal computer. Gateway utilized unique part numbers and lot-control for purchasing, manufacturing, and any warranty requirements.

When the Gateway team verbally reviewed the RFP content with the Siebel product team, it was something of a deer-in-the-headlights moment. The Siebel product engineer exclaimed: "We do not use a part number or bill of materials concept in Siebel. That just overcomplicates the sales and customer service processes. Instead, we leverage part descriptions to facilitate order entry and manufacturing." The Gateway engineering staff were dismayed. *Siebel used part descriptions instead of part numbers?* Without that level of granularity, it did not seem possible to meet any of Gateway's business requirements. In addition, Siebel could not meet approximately one-third of the remaining requirements in the same manner as the JD Edwards World software—the current working platform.

When the RFP team provided its findings to the Gateway leadership, along with the recommendation for the company to pass on the Siebel opportunity, the response was essentially, "We need Siebel in order to be relevant in the future. Figure out how to make it work." When the deal was finalized, it was reported that the software licensing cost $28 million. The acquisition was initiated on a golf course. It was rumored that the deal was signed and finalized there, as well.

Over the next two years, Gateway engaged two top-tier consulting firms to help "make it work." At the peak of the effort, seventy consultants were billing time on the Gateway/Siebel project. The cost to the Company was another $40 million in professional service fees. The Siebel platform finally went live for a portion of the Gateway sales team in 2002. It remained in production for a couple more years and was finally deprecated when the company's financial issues could not hide the inefficiency.

In conclusion, while an RFP can provide crucial insights for informed decision-making, it does not inherently ensure a good outcome. The Gateway-Siebel case serves as a poignant reminder that even with a thorough process misalignment in leadership priorities can occur.

CHAPTER 5 **SUMMARY**

- Multi-User Experience encapsulates any business relationship that is not with a customer or employee.
- Multi-User Experiences can be critical to the viability of a company's existence.
- Common multi-user stakeholders include regulatory agencies, banks, insurance providers, landlords, vendors, and all service providers.

- RFPs help to clarify your business needs and provide a standardized purchase analysis.
- RFPs can reduce business risk by identifying potential requirement gaps and mismatches.
- A formalized procurement process will not prevent management mistakes.

CHAPTER 6

Operational Excellence

In 1989, before Harry Stonecipher became president of McDonnell Douglas and Boeing, he accepted appointment as president and CEO of Sundstrand (now part of Raytheon). The company had roots in the furniture tooling industry and became an aerospace powerhouse during World War II. During the war, it was tasked with manufacturing constant-speed-drive units for aircraft. The hydraulic units converted the variable speed energy from the aircraft's engines into constant direct current electrical energy. At its peak in the 1980s, Sundstrand boasted of having components on every aircraft flying in the Western Hemisphere.

Stonecipher, a protégé of Jack Welch and ex-member of the General Electric management team, joined Sundstrand at a particularly challenging time. The company had just pleaded guilty to overcharging the U.S. Department of Defense on various military contracts. As part of a plea agreement, Sundstrand had agreed to pay $155 million in fines and contract reimbursements. Noteworthy, at the time it was the largest fine of this nature in United States history.

Aside from the litigation judgment, the company was facing tremendous financial challenges in inefficient manufacturing operations. Sundstrand had numerous component manufacturing facilities in the United States and Singapore. A costly fact—many of the products had to be sourced from multiple manufacturing sites. The inefficiency lay in the need to transfer products between facilities to complete manufacturing operations. For example, a component might start out in Denver, travel to York, Nebraska, for an operation, return to Denver for another operation, come back to York for final machining, and ship to Rockford, Illinois, for final assembly. To facilitate this activity, the company had a fleet of trucks. They were attractive semi-tractors with trailers sporting blue Sundstrand logos. Those trucks ran routes multiple times per day between the manufacturing facilities.

Stonecipher intuitively knew this inefficient practice was costing the company millions of dollars. When discussing the activity in one plant site visit, he exclaimed: "We are painting the damn highways blue with Sundstrand trucks! Do we even know how often parts are sourced complete in one manufacturing facility?" Amazingly, no one could answer the question. However, within a few months, the question was answered. At the time, only about 55 percent of the manufactured components for the company's aerospace segment was solely sourced in one component manufacturing center. The rest were cross-shipped between numerous sites.

You might be curious—how did the company come to have these dispersed manufacturing facilities? Rockford, Illinois, is no surprise, as the company was founded there. However, in the 1950s, Colorado offered tax benefits to build a manufacturing facility near the plutonium triggers production center near Denver called Rocky Flats. This incentive was complemented by the fact that many Sundstrand executives were falling in love with the winter sport of skiing. Further, many enjoyed the fly-fishing opportunities that Colorado had to offer. Downhill skiing proved to be the incentive for opening the Grand Junction, Colorado, manufacturing facility. Leadership found it much easier to get to Vail and Aspen by flying to the western side of the Continental Divide.

The electronics facility in Phoenix, Arizona, was "officially" built to take advantage of an excellent local workforce. However, every Rust Belt golfer knows you need to have a reason to visit Arizona in the January to March period. The same applies to the Brea, California, manufacturing center, as its high labor rate and product costs did not support its existence beyond its golf outing opportunities.

Sundstrand constructed its Singapore plant to take advantage of the country's extremely low labor rates, to establish an Asia-Pacific manufacturing/repair center, and to create offshore contract leverage with the Rockford, Illinois-based United Auto Workers union.

So, what about York, Nebraska? The area offers zero skiing opportunities, outside of the cross-country variety, and the golf courses are less than impressive. Well, the area does have one natural resource that abounds: pheasants. Unsurprisingly in the 1980s, Sundstrand executives loved to hunt pheasants. One senior leader, with relatives who farmed in Nebraska, promoted the idea of building a manufacturing facility there. Obviously, the state was thrilled to welcome a new complex into an area that had limited manufacturing industry. However, as a concession, the city of York and State of Nebraska had to agree to extend the local airport. This allowed for the Sundstrand corporate planes to more easily land and take off.

With the airport enabled and the new manufacturing facility about one mile away, it was not unusual for local Sundstrand employees to be sent to meet the planes at the local airport.

Often the purpose was not to retrieve corporate management or to receive expedited customer products; rather, they had to collect the bird dogs being shipped in ahead of the corporate pheasant hunts.

How Does This Example Depict the Need for Operational Excellence?

Whether your business is generating $10 thousand of revenue per year or $10 billion, you need to manage its operational effectiveness. This study of business philosophy is often referred to as *Operational Excellence*. This management philosophy promotes the constant search for ways to increase efficiency, improve product or service quality, reduce business costs, and/or generate greater customer satisfaction. To approach business excellence, you need to have relevant data for your business processes, employee performance, cost structure, and customer satisfaction.

Of course, no universal means of measuring Operational Excellence exists. We expect every business to view profitability or cash management as key indicators. However, Silicon Valley startups have often bucked that assumption by promoting company growth over profitability or cash management. The same odd behavior applies with customer satisfaction. Sadly, many companies view the volume of one-off sales or the success of their latest marketing campaign higher than the actual customer satisfaction scores with their products and services.

What Is Needed to Perform Data Analytics in a Repeatable Manner?

This discussion risks entering a "religious" argument between Microsoft's Power BI platform and Google's Looker Studio.

Business owners who embrace Microsoft for their word processing, spreadsheet, and various other administrative applications will find Power BI amazingly helpful. This platform enables you to build dashboards to analyze Total Experience-related data. However, Google proponents will argue that Looker Studio is the platform of choice. Both applications have pros and cons, but each is more than capable of performing analytics for businesses of all sizes.

Other platforms in the market include:

- Qlik Sense
- Sisense
- Tableau
- Zoho Analytics

Leveraging the spreadsheet applications like Excel or Google Sheets is perfectly acceptable for a start-up company. However, as your business grows, you will find it increasingly difficult to maintain daily, weekly, or monthly reporting needs with those tool sets.

What Makes Power BI and Looker Studio Analytics Better than a Spreadsheet?

One of the key differences is the ability to easily integrate different data sources to provide a visualization. With Google Sheets, you are limited to only integrating your spreadsheet with other spreadsheets. Excel has a little more flexibility to integrate other data sources, but they need to be small, or your processing time becomes exceptionally long. Microsoft's Power BI and Google's Looker Studio handle large and complex data sets efficiently. Another key difference is the ability to publish your reports or dashboards to members of your team or customers easily. You can achieve this in several different ways. For example, you might auto-generate reports and email them based on a distribution list or offer website access to a published dashboard.

How Can You Get Started with Operational Excellence?

Numerous books dive deep into the philosophy and application of Operational Excellence programs. You will find many of the more popular selections listed in the Bibliography. Those selections lean toward the Toyota Production System popular in the manufacturing sector. However, Figure 6.1 depicts the handful of basic steps needed to begin an Operational Excellence effort for any business.

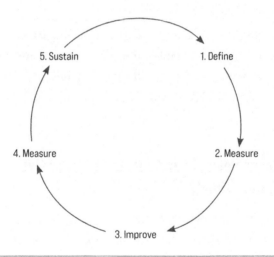

FIGURE 6.1 The required steps to begin an Operational Excellence effort

Begin by defining the problem or process you want to improve, determine a means to measure the current performance, identify a means to improve it, measure your success, and if the results are favorable, ensure the process is sustained.

Harry Stonecipher emphasized the need to fully understand the business's manufacturing processes and its associated cost structure. He argued the only way this could be accomplished was via the religious use of data. He used data to make decisions to improve Sundstrand's profitability and to enhance the company's future viability. Given Stonecipher's data-driven decision-making process, Sundstrand would no longer "paint the highways blue with Sundstrand trucks."

Applying this approach to your own business involves delving into the nuances of your operations, ensuring that every aspect is informed by accurate and up-to-date data. This commitment to data-driven decision-making can be transformative. It will enhance your current operations and position your business for sustained success in the future.

CHAPTER 6 SUMMARY

- Operational Excellence means striving for continuous improvement in all areas of your business and its processes.
- Operational Excellence by its definition requires a means of assessing the effectiveness of your business processes.
- Operational Excellence requires data to benchmark between acceptable or unacceptable outcomes resulting from a business process or activity.
- Data analytics platforms can be extremely helpful in aiding your business reporting to customers, employees, and other stakeholders.

CHAPTER 7

How Do You Get Started in Total Experience?

Mark Twain once said: "The first step is always the hardest." A similar sentiment, attributed to author Steve Pavlina with respect to successful projects, emphasizes that "the toughest act of completing a project is getting started." Both thoughts apply to starting Total Experience efforts. However, it is not a "one and done" endeavor. Rather, successful business leaders will consistently evaluate the organization's Total Experience. They will always be looking for ways to improve it.

The following outlines two series of steps to help you start your Total Experience journey. To avoid getting stuck in analysis paralysis, a good approach is to dedicate a few hours to the first series and up to a few days on the second series. Given the potential shifts in your business environment over the course of a year, make a point of revisiting the first series of steps on an annual basis.

Series One:

- Define your business model and set your Total Experience goals.
- Define your stakeholders in the Total Experience model (for example, new clients, existing clients, employees, or "x" regulatory agency).
- Define your ideal customer, employee, and multi-user journeys.

If you do not have clarity on these three steps, you are not ready to try to fix anything. It is equivalent to starting a car journey but not knowing where you are going, who will help you with your journey, or how to define a successful trip.

Understanding your stakeholders is especially important. Your customers and employees always come to mind. However, if you are not also considering your product vendors, bank lender, insurance broker, and such, you will miss some key components of your Total Experience.

Series Two:

- Identify and prioritize your existing pain points based on business impact.
- Collect data on the processes to measure success.
- Identify the services/technology applications that can minimize your top pain point or completely resolve the issue.
- Assess the viable solutions for cost/benefit, time to implement, and likelihood of success.
- Choose the most value-added option to implement.

- Engage your employees to help implement the solution—obtain their support.
- Solicit Total Experience stakeholder feedback (clients, employees, and others).
- Adjust the process where necessary and reassess your success against your goals.
- Choose the next pain point and attack it in the same manner ("rinse and repeat").

This pain-point remediation approach follows the guidance that the economist and quality-control expert W. Edwards Deming provided in his classic book, *Quality, Productivity and Competitive Position* (1982). Deming was blunt, "It is wrong to automate a bad process. It is even worse to automate it twice. Never automate a process without first improving it."

There is a caveat to looking at the list of pain points and solving for one. It is possible that you may be able to address related issues concurrently by the same solution. Returning to Maggie's office chaos, her hospital clients and her nurses were struggling to contact her for schedule changes. By implementing a voice Customer Experience solution and assigning specific telephone numbers to existing clients (another one advertised for new clients and a third for her nurses), she could more easily prioritize the operational inbound support of all three groups. Going one step further, she could introduce an Interactive Voice Response answering attendant that might provide more options for self-service. This could include electronically authenticating the callers and performing a selectable function (for example, nurses call in to validate their next assignment time or determine the address of their clients). Those types of text-to-speech readouts have become easy to automate.

Prioritize Your Pain Points

A common mistake many companies make is trying to fix everything at the same time. They use the shotgun approach to improve experiences by starting numerous initiatives all at once. The hope is that one or two of them might finish successfully, and the other activities will not make things worse. The core issue with that approach is limited resources. Trying to do many initiatives all at the same time limits your ability to focus and achieve one positive result. The other issue is having the ability to effectively measure all the changes with certainty of cause and effect. If you trigger ten changes all at once, without knowing which is providing the favorable result you are looking for, there is a good chance that a future change might negate the benefit.

Collect Data to Measure Success

Data collection can be as easy as recording the occurrence of an event on a scratch pad each time it happens. After a period, aggregating the events into occurrences per day (or another appropriate period) allows you to report on a trend.

This baseline tells you how bad the process is per day, per week, or whatever. Thereafter, when you try to implement a remediation, you can quickly see if it is helping reduce the pain point and by how much. Your first solution may need to be adjusted. By continuing to collect data on the events, you can monitor the data to ensure the pain does not return.

Issue Resolution Options

This is where talented process consultants prove their value. Alternatively, this activity proves that a little bit of internet research can pay huge dividends. Once you understand your biggest pain point you need to identify your options to resolve it. Or, at a minimum, improve it. This is where franchisees tend to have a bit of an edge over their independent small business peers. Most franchisors have business/process consultants on staff to aid in process engineering. That effort and the cost of usage is typically captured within the franchise fees. Further, the network of other franchisees is an excellent resource for problem-solving. Independent business owners typically must either purchase business consulting or do the research on their own.

Assess the Viable Solutions for Cost/Benefit, Time to Implement, and Likelihood of Success

The toughest aspect of identifying and selecting options to resolve the issue is getting holistic cost estimates for each option. If the preferred approach is purchasing a software application, you can get the subscription or license cost easily enough. However, the implementation time and effort are usually harder to put a price tag to. The goal is to avoid spending $1,000 to solve a $100 problem.

This is where a business consultant can help quantify the costs and the implementation effort. If you do not have access to a trusted advisor, you can usually find a good amount of helpful information in business consultants' blog articles about the various software vendors and applications you may be considering.

Choose the Most Value-Added Option to Implement

Recommendation: try to pilot or model the solution as simply as possible before purchasing subscriptions to software applications. The example we covered on Workforce Management applies here. If you need a teammate scheduling tool, try leveraging the spreadsheet model first. If you can initiate a workable solution without purchasing software and still significantly reduce business pain, great! If not, you have gained the learning of what is important to have in your workforce planning application to meet your business needs.

Of note, many software vendors will allow you to do a proof-of-concept installation via a thirty-day free trial (or some other meaningful period). A strong recommendation is to push for a proof-of-concept period when talking with software sales reps.

Engage Your Employees to Help Implement the Solution— Obtain Their Support

It is amazing when business leaders choose a new path and forget to prepare their employees for the journey. The harsh truth: you can identify a perfect solution, but if your employees do not support it, the implementation will fail.

One of the quickest ways to achieve employee buy-in is by encouraging participation in choosing the new process method or software application. If that is not possible, then the next best option is to explain: (a) why a change is necessary, (b) why you have chosen this approach, and (c) how you expect the new process to help them do their jobs.

This last tenet assumes the new process/application will help them perform their jobs. Hopefully, that is the case. However, if it creates an added employee burden but benefits a third party important to the organization (like a regulatory agency), it can still be successful. You just need to double-down in communicating why the new process is important to the business.

Once you have the engagement of the employees, the next thing is to gain agreement on the success criteria. This should reference back to the data you are collecting to quantify the pain point. After you have laid the foundation for success with the employees and know what you want to achieve, the last step is to implement the change. The best approach is to perform a small pilot implementation. However, there are times when a "Big Bang" implementation is required. If that is the case, you should have a rollback strategy; meaning: "If this change isn't working, how can I return as quickly as possible to my prior process?"

Solicit Total Experience Stakeholder Feedback

Hopefully, your new process is a resounding success. You compare the data collected against your success criteria and see the pain point reducing in size. However, this is where soliciting feedback from your customers, employees, and other stakeholders is a critical step.

Your new process may help one aspect of the business, but it may have created a new, even larger pain point that no one anticipated. Some leaders assume that clients will automatically tell them if a new issue has been introduced. However, that is not always the case. Often, they will try to live with the new condition for a period—only speaking out when the pain becomes too severe. Asking for immediate feedback eliminates the risk of incurring an unintended consequence that can cost your organization clients and revenue.

Adjust the Process and Reassess Your Success against Your Goals

This is the basis of continuous improvement. You have implemented the new process. It is working OK. By tweaking it slightly, you think you can make the results even better. Do it and reassess. If you find that the change is not immediately helping, consider reverting back.

This seems like common sense. However, many leaders look at a change as a one-time event. They miss tremendous opportunities to achieve even better business results while the focus on the newly implemented process is fresh in everyone's mind.

Choose the Next Pain Point and Attack It in the Same Manner

When you are convinced that you have your top pain point satisfactorily resolved, then move on to the next one. This is the "rinse and repeat" process for achieving Operational Excellence.

This approach to resolving pain points assumes that your business has been running for some period and pain points are regularly occurring events. It is a sound methodology to pursue.

However, there is a compelling argument for establishing best-practice methods of operation *before* launching the business. For example, many new restaurant owners will schedule several "friends and family" nights with this concept in mind. During these events, staff members invite friends and family members to enjoy a free meal hosted by the business management. These popular—and strategic—events allow the company to test their recipes and iron out service delivery practices without the risk of customer complaint.

Similarly, most franchisors will require their new franchise owners to attend training classes or spend time in an existing franchise. They use this method to teach the basic processes needed to operate their start-up businesses. With this approach, you are designing the launch of your business with Total Experience in mind. This planning increases the odds of new business survival. The International Franchise Association published a report that found the five-year survival rate for new franchises is 92 percent versus a 50 percent survival rate for independent businesses. That suggests this learn-before-launch approach holds tremendous value in establishing good Total Experience.

CHAPTER 7 SUMMARY

- Total Experience is an ongoing process requiring continuous evaluation.
- Total Experience requires business owners and leaders to:
  - Understand their business model.
  - Identify key stakeholders in the business performance.
  - Understand their ideal customer journey.
- Business owners and leaders can strive for Total Experience by systematically reducing business pain points for customers, employees, and other key stakeholders.
- For a Total Experience improvement program to be successful, you must achieve employee engagement.

CHAPTER 8

Key Performance Indicators Needed for Success

In 2019, a franchise corporation was enjoying a wonderful period of continuous growth. For over a decade, the company had participated in respectable year-over-year revenue growth. Serviced client numbers were on the rise, and annual measurements of employee sentiment consistently reflected positively. The overall health of the business was viewed as strong.

The following year, the COVID-19 pandemic began, which stressed the business model. While the franchise network's revenue continued to increase, client acquisition plateaued. Though leadership expressed concern, it was not until 2021, amid the ongoing pandemic challenges and additional business headwinds, that the corporation's leaders observed a decline in serviced client numbers. Surprisingly, year-over-year revenue growth still seemed favorable. All appeared well, or so it seemed.

In 2022, the tide had turned. Revenue growth halted, and serviced client counts were now in significant decline. Corporate leadership, sensing a deeper issue, sought more data to understand the true business environment. The ignored metric was the "hours of service being provided."

Upon incorporating this business indicator, corporate leadership discovered a disconcerting reality. After the pandemic, the franchise owners had increased service pricing to maintain revenue. While this strategy initially maintained local office revenue (and increased profits by reducing office costs needed to service fewer clients), it was proving disastrous for the franchise network's overall growth. The franchise brand and business was shrinking and losing market share.

Key Performance Indicators (KPIs) are the handful of measurements you will use to assess your Total Experience success. In the pre-pandemic scenario for this company, KPIs focused on revenue, client growth, and employee sentiment. For many years, those indicators serviced the company very well. However, economic shifts prompted a change in the key indicators, and they added an "hours of service" metric to gain a clearer understanding of the actual situation. Although various metrics could have identified the source of the unfavorable trend, the focus on "hours of service" offered valuable insights for this particular company.

Every business should prioritize basic financial indicators: revenue, net income, and cash flow. However, the additional business metrics needed for your organization are likely to be unique to your business model and industry. As a rule, your KPIs should help you:

- Establish clear business goals.
- Track your progress against those business goals.
- Identify new business opportunities.

For example, a manufacturing business may track raw material inventory that it has on hand and purchase-to-delivery lead time. If your business is in the real estate industry, you may want to track your listing days on the market and client referrals.

Automotive dealerships might analyze new car sales versus used car sales and the average gross profit per unit sale.

The key takeaway is that KPIs are essential for measuring Total Experience performance, and these indicators need to align with your business model. They may also need to change depending on where you are in your business growth cycle and the current economic environment.

The eBay example in Chapter 2 illustrates the importance of specific customer indicators like the Net Promoter Score. Another indicator, unique to eBay, was tied to the marketplace feedback mechanism built into its website. If leaders saw a significant uptick in negative scores for a particular product segment, they would dig deeply to understand why it was happening. Most brick-and-mortar retail stores do not have that customer feedback system available. Rather, they monitor the amount of merchandise customers are returning to indicate if they have a product quality or sourcing issue.

Figure 8.1 shows a few of the key areas that most of the books discussing KPIs recommend as important.

Financial Perspective	Customer Perspective
• Revenue growth • Profitability • Return on investment (ROI) • Cash flow • Market share	• Customer satisfaction • Customer retention • Customer loyalty • Market share • Brand awareness
Internal Process Perspective	Learning & Growth Perspective
• Process efficiency • Cycle time • Quality • Productivity • Cost reduction	• Employee satisfaction • Employee retention • Employee training and development • Innovation

FIGURE 8.1 The most important areas to focus on when tracking Key Performance Indicators (KPIs)

If you want to dive deeper into common KPIs and how to measure them, the following book is a highly recommended text: *Key Performance Indicators (KPI): The 75 Measures Every Manager Needs to Know* by Bernard Marr (2012). Within it, he dives into many of the most common KPIs covering finance, customer feedback, marketing, sales, manufacturing, employee feedback, and even corporate social metrics. The book is structured to allow you to focus on indicators that would fit your specific business.

Establishing meaningful KPIs to measure Total Experience is another area where business consultants can offer valuable assistance. Franchisors, too, typically provide a set of KPIs during training to enhance franchise owners' chances for success. For start-up business owners who crafted a business plan for financing, revisiting that document and basing initial KPIs on the projected metrics can be beneficial.

To effectively address business pain points, you will need data, and KPIs are essential to ensure that actions yield the desired positive impact. Ideally, these KPIs become part of your ongoing portfolio of metrics. Alternatively, you can aggregate detailed metrics into a macro KPI for general monitoring. For instance, if your business faced the challenge of clients dropping calls due to long wait times, you might have monitored the average on-hold wait time to resolve the issue. However, this level of granularity may not be suitable for monthly business KPIs. Instead, you might opt to measure new client adoption rates. If this rate begins a negative trend, a review of several detailed metrics, including new client average on-hold wait time, can help identify the root cause of the change and allow corrective action.

CHAPTER 8 SUMMARY

- KPIs are critical to measure the performance of your business against your business goals.
- KPIs can help you identify issues before they become major problems affecting your customers, employees, or business health.
- KPIs may help you identify new business opportunities.
- KPIs vary significantly between industries, business models, and the stages of the business life cycle.

CHAPTER 9

Total Experience Staffing

Many business owners struggle with the decision: "When should I hire additional staff to address Total Experience initiatives, and what roles should I target?" While it is impossible to provide a one-size-fits-all answer to this question, you should consider adding Total Experience staff when the benefits of increased productivity, improved customer service, or business growth outweigh the increased cost of the resources. Also, consider adding Total Experience staff when the risk of losing existing business due to increasing customer needs outweighs the costs and challenges associated with onboarding a new employee.

The core premise is not a revolutionary business concept. It has been the basis by which small business owners have addressed staffing changes since the dawn of capitalism. However, the new difference is adding the concept of Total Experience to the calculus. With it included, you are considering all business stakeholders in the equation, not just customer sentiment.

Some staffing decisions are easy to make. Looking back at Maggie's nursing business, she identified the need for more nurses to satisfy her growing customer demand. However, it was tougher to assess the office staff requirements. She had two administration teammates on board to help address nurse scheduling, phone support, billing, and other office duties. Theoretically, Maggie could keep adding physical office employees to address the problem. But there is a point when manual handoffs of business activity become error-ridden and horribly inefficient. When your business is at that point, adding more staff makes office chaos worse.

If Maggie's only issue is telephone coverage, she has multiple options. She might be able to subscribe to an answering service to handle inbound telephone contacts, take messages, and answer basic questions. However, if the condition is across the breadth of her business, she may need to automate functions with software. If she chooses the software route, Maggie will need someone with special skill sets (either part-time or full-time) to help digitize the business processes. By automating activities for her customers and nursing staff, her two office administrators should become more efficient.

Owners who have made those timely decisions often see improved Total Experience across other areas of their business. With more time, office teammates can address other beneficial functions that are often ignored like quality check-ins with clients, employee feedback surveys, and improved relationships with multi-users such as regulatory agencies.

If Timely Decisions are Important for Growth, How Can You Make the Best Staffing Decision?

Networking is widely acknowledged as a key to personal success, and this especially holds true in small business staffing activities. Business owners and leaders who forge connections with other business leaders or advisors tend to fare better than those who try to go it alone. Trusted advisors, having encountered similar decisions, are willing to share their experiences. Most are equally willing to disclose their decisions and the resulting business outcomes. Truly valuable advisors will share their mistakes as well as their successes. Moreover, those who offer candidate recommendations with the required skill sets are invaluable. Leaders agree that candidate referrals consistently yield the best new employees.

Furthering this train of thought, I recently collaborated with a franchise network that had instituted a "virtual board of directors" concept among subsets of the network's franchise owners. The program grouped eight or more owners together. This pseudo board of directors would meet two or three times a year. During those events, they reviewed each other's common business metrics and associated staffing challenges. The owners typically managed businesses of comparable size and complexity. Although the staffing models evolved based on business size, some successful owners in these "board of director settings" opted to blend Total Experience roles. For instance, some leaders combined the roles of recruiter and scheduler into one job. This approach allowed them to distribute the workload among multiple staff members, segmenting tasks based on factors like client location. The underlying benefit was having trained backups for business functions across multiple office members. Consequently, when one staffing recruiter/scheduler had to be out of the office for a few days, others were ready to fill in, ensuring critical functions did not come to a halt.

The staffing consideration above introduces one more key aspect to Total Experience. When making sourcing decisions, you need to be mindful of resource gaps while maximizing day-to-day efficiencies. In other words, leaders should be cautious not to rely solely on one employee for a particular function. This often occurs when an employee is exceptionally efficient or possessive of a particular task. This might be a business necessity in the short term. However, for the long term, having multiple employees cross-trained in all business workflows is essential.

A lender of short-term loans shared an example of a Total Experience failure that involved a client sourcing a workflow to only one key employee. One of the lender's bridge loan customers had organized specific work activities exclusively with specific staff members. The employees' roles included a client coordinator, a scheduler, a procurement specialist, and a few other roles uniquely assigned—with no cross-training. One of those key roles was invoice billing. Unfortunately, that individual decided to leave the company. For six weeks, no customer invoicing took place. The business owner quickly realized the gap when trying to secure another bridge loan to cover payroll expenses. Invoice documentation was unavailable to support the lending event. As you can imagine, this made processing the next payroll an incredibly challenging exercise.

In the realm of staffing, another Total Experience consideration is establishing the right culture within your business. This involves various facets. Identifying the right recruits based on skill sets and attitudes is critical. However, demonstrating excellent leadership principles from the start of employment is equally critical.

Here is a prime example of a leadership role model. Before hiring a new finance director, Tom Hoag, a divisional CFO for IDEX Corporation, arranged for the candidate to undergo a series of interviews with himself and the rest of the division leadership. By the time the company extended the job offer, the candidate and the leadership team were well acquainted. Even so, on the recruit's first day in the office, Hoag made a point of spending the first hour of employment together with the recruit. He used the time to describe his leadership style and expectations. For example, he noted that he was often the first employee in the office, being an early riser, but that he did not expect his team to arrive ahead of normal business hours. He highlighted existing company culture items, shared a new org chart, including the new employee's role, and described the team goals for that year. Near the end of the dialogue, after addressing job goals, he proclaimed: "I welcome healthy debate on pending decisions or actions. However, I will not tolerate disrespect—in public or in private."

Wow! Within the first hour of employment, the new employee had five pages of notes on the company and team goals. He understood his place in the organization. He knew his initial job goals and how to avoid infringing upon sacred company "culture elements." Lastly, the recruit clearly understood how much latitude he had to challenge his boss's thoughts. No doubt, Hoag had spent more than a few hours preparing for that meeting despite his busy schedule. But he prioritized getting his new employee off to the right start.

That leadership example was not a one-off event. Hoag "walked the talk" with his leadership style and expectations. He was willing to discuss any pending decisions with an open mind and could be swayed to change his position. However, once the decision was made—unless external factors affected the calculus—he expected the effort to be delivered accordingly.

Compare that experience to the following new-hire onboarding story with a different company and leadership personality. The hiring manager spent three months making up his mind to extend the job offer (red flag). However, upon making the offer, the area leader pushed to delay the start date an additional four weeks; this was supposed to better align his schedule with the entry into the company. On the first day of employment, the leader was only able to support a thirty-minute conversation with the recruit. During the discussion, he spent his time describing a pending project that would be part of the job. The rest of the recruit's day was spent with the human resource team and completed onboarding rituals.

On the second day, the new hire was invited to attend a senior leadership meeting where an organization chart was presented. Interestingly, his role was significantly different from that discussed during the interview process. Throughout the following two weeks, the hiring manager was able to provide another sixty minutes of time to discuss job expectations, staffing, and organizational structure. Unfortunately, only vague answers were forthcoming to questions. When the leader was asked about differences between the description of the job during the interview process and what was now materializing, his response was, "We never said this role was going to be easy."

Leadership matters. Whether it is Day 1 or Day 720, you need to bring your best management effort and integrity to the business every day. If you want employees to maximize their efforts toward Total Experience, you need to lead by example.

Reflecting on companies in the news with outstanding business achievements (Apple, Pinterest, ServiceNow, Tesla, and so forth), their leadership brings their A-game to every employee engagement. Conversely, organizations with high employee turnover rates, declining customer engagement, and vendor management issues tend to have leadership that is often not "walking the talk."

In the realm of recruitment, many leaders promote hiring for a customer-centric mindset. Businesses adopting this approach are likely to outperform those who hire teammates indifferent to Customer Experience. This was PayPal's approach prior to 2020. The company vetted customer service teammates who were aligned with Total Experience needs. They wanted teammates who genuinely were committed to easing customer frustration and displaying empathy. Additionally, these teammates were expected to welcome empowerment to make good customer account decisions. This necessitated finding new recruits who were concerned about the overall health of the company, who had a history of good judgment, and who demonstrated a passion for fairness. Employees understood the goal was to achieve a fair result for both the customer and the company. This culture of trust yielded low attrition rates and encouraged employee referrals of friends and relatives for open roles.

PayPal trained its customer service teammates in the basic elements of continuous improvement. The company encouraged them to consistently identify inefficient workflows or product bugs. Their incentive was not purely monetary; rather, it was based on a desire to reduce future customer frustration and contacts.

What If Your Business Is Not Growing, but Shrinking?

As mentioned in this book's Introduction, the focus of the text is addressing business growing pains versus assessing downsizing options. However, when your Total Experience assessment is telling you that your market has changed or business conditions demand a staffing adjustment, doing so with a mindset on Operational Excellence is prudent. If you are forced to reduce staff to balance your cost inputs, choosing those roles that minimize the impact on Customer Experience and overall Operational Excellence is basic common sense. A book that dives deeper into how to perform downsizing assessments is *How to Downsize Your Business Successfully* (Entrepreneur Today Series Book 8) by Mark Patene (2014-15).

How Can You Best Influence Total Experience through Staffing?

Fostering a Total Experience culture begins with leadership. Recruiting individuals with the right mindset is critical. The management focus extends through the onboarding process and never ends. Leaders must encourage a passion for continuous improvement in all business operations. They must also exemplify it themselves.

CHAPTER 9 SUMMARY

- Timely staffing decisions are optimized by considering Total Experience implications.
- Total Experience staffing principal: Add staff when new benefits outweigh the costs.
- Networking with other business leaders can help with staffing decisions and recruiting.
- Leadership matters when striving for business optimization.
- The goal is to strive for a Total Experience culture that maximizes customer and employee satisfaction in concert with business value.

CHAPTER 10

Total Experience Mistakes to Avoid

Almost everyone over the age of twenty has heard of Motorola. While it was a more than fifty-year-old company and a household name by 1980, it did not hit its global stride until the cellular phone industry exploded later that decade. Everyone was buying Motorola cell phones. The company had numerous other technology divisions supporting tremendous corporate growth. However, in the 1990s the Motorola leadership viewed the corporate performance of Toyota with envy. They were convinced that Toyota's lean manufacturing practices and Total Quality Management (TQM) approach were the recipe for future success. TQM was marketed as the holy grail for excellent corporate performance and longevity. If your products and processes were nearing perfection—having zero defects—you could be sure your customers would love you, and brand loyalty would be unshakable.

My colleagues who worked for the company in the 1990s highlighted a dangerous habit. Motorola's leadership focused so much energy on TQM that it often overshadowed new product development. Specifically, spending resources to gather and analyze product quality data was often multiples higher than the cost inputs to design, manufacture, and distribute the analyzed products. This imbalance created a sideline effect. The company spent so much focus perfecting aged technology that it was being left behind by competitors innovating new products. This was proven the case in 1998 when Nokia overtook Motorola as the world's largest cell phone producer. The introduction of the iPhone in 2007 by Apple all but eliminated Motorola from the cell phone landscape.

This is a prime example of a company getting too wrapped up in its business enhancement programs. It did not maintain a good balance between innovation, cost management, product delivery, and customer satisfaction. Motorola was spending mountains of cash to obtain the last two percent of product perfection when, in 2007, customers did not value having the "perfect 1990s cell phone." They certainly were not going to pay a premium for a zero-defect product that lacked technological innovation.

Besides the risk of over-indexing on specific Total Experience programs, business leaders need to be cautious about getting "oversold." Small business owners tend to be susceptible of being oversold on service offerings for recruiting, billing functions, and Workforce Management applications. These are not typically gullible professionals. However, business pain points can be overamplified by aggressive sales associates.

A good example was a business owner who was struggling with the intricacies of government-funded payments. The billing process was quite complex. Any missteps in the invoice submission or services coding allowed the government to defer payment while the errors were corrected. These deferrals often stretched over many months. Although government contracts did not form a substantial portion of the owner's sales, they constituted the most labor-intensive segment to administer within his book of business. Unsurprisingly, it also represented the highest accounts receivable balance.

Enter a sales representative from a specializing government billing company. Persuaded by the promise of streamlined processes and reasonable fees, the owner decided to process the company's invoicing through this external organization. It was a commendable move forward. However, the salesperson then convinced the owner to leverage the company's prepayment of the receivables. Obviously, this was a loan structure. A closer look exposed the excessive cost of this financial arrangement, it included an effective interest rate of 26 percent. Within a few months, the business owner realized that the convenience of the prepayment process was devouring all his profits derived from that business segment.

Which Business Segment Tends to Avoid Total Experience Mistakes?

While no segment is immune to Total Experience mishaps, one group that tends to be commendable is farmers. They typically have low-margin businesses. Their markets are susceptible to tremendous price swings that may be driven by weather, geopolitical, economic, or technological changes out of their control. The one thing all farmers can easily articulate is their business goals. Most can also quickly explain their strategy to achieve those goals. They want to maximize their production yields, control their cost inputs, and ensure good stewardship of their land and animals, so the farm is viable for future generations.

Additionally, most farmers leverage old-fashioned, common sense when approaching business challenges. For example, when comparing employees to crops, you might hear a farmer advise:

- Do not shout at the crops.
- Choose the best plants for your type of soil.
- Weed your fields.
- Do not blame the crops for not growing fast enough.
- Help the crops to grow with resources versus running over them.

Consequently, when my grandfather—a third-generation farmer—gave me early career advice about being "penny-wise but pound-foolish," I took it to heart.

During the 2021 wage inflation period, I was collaborating with a small business owner who was grappling with a labor supply problem. He was holding firm to a labor rate of $11 per hour for his service-providing workforce. This average rate was a figure frozen in time for four years. Despite a robust customer waiting list, labor scarcity was hindering business expansion. Adding to his woes, his business was suffering from high employee attrition and very significant process cost to hire fresh staff members. This process expense included the pain of coordinating interviews, vetting the individuals, completing background checks, and performing new-hire training. Adding insult to injury, his tenured staff was tired of working overtime hours and was growing increasingly disgruntled.

I asked him how much he was spending in time and fees to recruit one new employee. He estimated each recruit was costing his business $2,850 to obtain and train. I learned that he was losing six employees each month to attrition. The follow-on question was: "How much could you increase your labor rate if you reduced your attrition rate by 50 percent?" He was not sure, so we performed some "back of the napkin" calculations. Those calculations are detailed in Figure 10.1.

Cost to hire one new service worker [a]	$	2,850
Average employee attrition per month [b]		6
Total monthly attrition replacement cost [a * b]	$	17,100
Attrition savings at 50% [a * b] * 0.5	$	8,550
Total service workers [d]		65
Average hours per week per employee [e]		20
Weeks per month [f]		4
Total hours per month [d * e * f]		5,200
Total labor cost @ $11/hour [d * e * f] * 11	$	57,200
Potential rate increase from reducting attition by 50% (total savings/total hrs)	$	1.64
Existing average labor rate for service workers	$	11.00
New potential labor rate (existing rate + savings increase)	$	12.64

FIGURE 10.1 An example that proves it is often more cost effective to raise the hourly wage to retain employees to significantly reduce the costs and business impacts of high attrition

Without trying to get fancy by including the cost of high overtime, client frustration with the employee churn, and so forth, we quickly concluded that he could raise his average labor rate from $11 per hour to nearly $13 per hour, which could favorably impact attrition.

As the conversation continued, I asked how many of his waiting list clients would pay an extra $5 per hour for his services. The idea was to pass a small price increase to new customers, increase his average labor rate for his entire workforce, and promote an even larger favorable impact on his labor attrition. Because he had a significant waiting list of new clients, he felt it was worth a try. After raising his average labor rate to $14 per hour, he reported the impact on his business was impressive. His employee attrition was nearly eliminated, and his business was now growing at an annualized 12 percent rate.

The point of this example is to highlight the risk of being "penny-wise but pound-foolish." By being reluctant to address the wage pain point, the business was generating a fly-wheel effect of employee attrition, recruit, train, and labor erosion. To avoid mistakes like this, you must have a good handle on your business expenses, your employee feedback, your customer sentiment, and the overall market. Therefore, think like a farmer: invest wisely, monitor results, and adjust when needed.

What about the Management of KPIs?

Building on the need to measure and track Key Performance Indicators, as discussed in the previous chapter, it is critical to avoid a common misstep in their application. Strangely enough, businesses often adopt a committee-based approach to KPI accountability rather than assigning a specific individual responsibility for each indicator. The rationale often echoes the sentiment: "We value teamwork above all else." While fostering teamwork is commendable, relying on a committee approach in this area is not the optimal path to Operational Excellence. When multiple individuals perform the same function, it is important to have a means to identify individual performance.

In addition, if you attempt to hold a staff member accountable for a peripheral impact on a KPI, you diminish the KPI's significance and risk losing the employee's respect. Moreover, the individual truly responsible is unlikely to change his or her behavior. Therefore, the problem will often persist and be construed as unfair to the rest of the team.

Consider a scenario where a retail business relies on timely submission of new purchase orders to replenish inventory. A metric tracks the late creation of purchase orders, pointing to the root cause of delayed inventory deliveries. Now, imagine having a procurement agent responsible for purchasing new inventory and an accountant responsible for paying supplier invoices. If coaching efforts focus on holding the accountant responsible for the late POs, delayed shipments, and empty shelves, it is inherently unfair. The coaching opportunity should squarely address the procurement agent, who directly influences the issue at hand.

What Happens If You Rely on Stale Total Experience Data?

Regrettably, using outdated data for critical business decisions is one of the most common errors made by businesses of all sizes. This oversight can manifest when leaders base their decisions on customer, employee, or vendor feedback that has become obsolete.

A case in point involves a large company that routinely conducted employee feedback surveys every April to assess sentiments regarding management decisions, behavior, and other business considerations. In one particular year, the leaders decided to proceed with the April survey while knowing a major company restructuring was pending. In May, the reorganization included employee layoffs. After the event, the company published the April survey results. Team leaders reassured executive staff on employee morale, citing the anonymous employee responses from the annual survey. Managers reported that employee survey feedback reflected a small number of employees who were considering leaving the company, but overall trust in management was rated as high. Leaders were shocked when voluntary attrition rates more than doubled two months after the layoffs.

How Do Mergers and Acquisitions Affect Total Experience?

This topic is worthy of a book unto itself. Mergers and acquisitions tend to be tremendously disruptive to both the acquiring and acquired parties. However, companies can follow a few guidelines to minimize the disruption—both in the short term and long term.

If the two corporate entities are in the same industry and anticipate sharing customers, a good rule of thumb is to complete the systems and process integration as quickly as possible. You can find numerous examples of companies that tried to "leave the acquired entity managed separately."

When the two companies are in different industries, with different customers and different product or service offerings, maintaining separate operations makes a great deal of sense. Berkshire Hathaway Inc. has been highly successful with this model. However, PayPal's experience with the purchase and integration of Bill Me Later, Inc. is a poster-child example of delaying the integration too long. It took many years to complete the integration due more to corporate politics and egos than to any technological impediments. This generated tremendous costs and a wasted opportunity with their shared customer base.

IDEX, Inc. set the gold standard for evaluating acquisitions, communicating post-acquisition plans to the acquired leadership, and then completing the consolidation of systems and processes. The conglomerate is based upon the micro-acquisitions of more than 100 companies over the last thirty years. Many of its companies overlap between products and customers (for example, Viking Pump and ABEL Pump Technology). However, many of the companies, within their three business segments, support distinct customers and would not benefit from full standardization across the enterprise.

In the early 2000s, IDEX recognized the potential for cost savings by completing as much technology and process standardization as possible. Its corporate success (and stock price performance) over the last twenty-five years is a testament to that corporate leadership philosophy and the company's best-in-class models for strategy execution.

Conversely, a building materials behemoth has been extremely focused on acquisitions for several years. This corporation acquired many competing companies, sharing many of the same customers with remarkably comparable products. The leadership placed standardization of processes and technology so low on the corporate priority list that the ability to report financials became a struggle. New leadership has recognized the need for Total Experience improvement across the corporate landscape. However, delays in strategy execution have turned a complex problem into an execution nightmare. This includes end-of-life systems, manufacturing inefficiencies, leadership egos, and a "Let's just throw money at the problem" mentality. Where IDEX has proven to be able to integrate a company and its associated manufacturing entities in six months, this building supply company has taken three years to integrate one manufacturing entity for one division.

Another consideration is choosing the best technology for the combined entity's standard offering. In most cases, the acquiring company's technology profile is adopted. The acquirer typically has a larger footprint of corporate assets needing to migrate, and deems it less costly to convert the acquired entity to the existing corporate standard. However, there have been times when this is not the case. Acqui-purchases (targeting both the acquirer's technology and staff) may make it more appropriate to standardize with the acquired company's technology stack. Either way, it is worth doing an analysis of the pros and cons for both scenarios. Leaving corporate egos at the door when making a value-based decision is critical for business success.

Understand the Business Problem Before Selecting a Solution

A frequent business challenge is addressing leadership passion for a specific application or service without understanding the business problem needing to be solved. Numerous examples of this exist across every industry. However, one exchange between a new company president and the tenured CIO regarding a reporting engine went something like this:

President: "I can't believe we are not reporting on this business metric! The Acme reporting engine that we used at my last company was awesome! All our customers loved it! We need to buy the Acme software here!"

CIO: "The issue is not that our existing reporting platform cannot provide the same report as the Acme engine. Rather, it is the fact that the clients do not provide the data elements needed to create the report."

President: "OK, so how long will it take to swap out the reporting engine with Acme?"

Opportunities to misjudge Total Experience actions abound for every business. The example above exemplifies situations where the final desired outcome is understood, but the business constraints are not commonly recognized.

Decisions affecting business change can be difficult. However, in today's competitive landscape, the cost of doing nothing in support of Total Experience can be catastrophic. Creating positive change is every business leader's key assignment. A favorite quote by John Augustus Shedd captures this thought: "A ship in harbor is safe, but that is not what ships are built for" (1928).

CHAPTER 10 **SUMMARY**

- Over-indexing on a specific Total Experience area can be detrimental to business performance.
- Be cautious about buying more products and services than are needed to address a Total Experience issue.
- Business owners and leaders need to define clear accountability for each significant KPI.
- The use of stale data to make decisions affecting Total Experience is a frequent problem.
- Understand the business problem before selecting a solution.

CHAPTER 11

Artificial Intelligence and Total Experience

Numerous artificial intelligence (AI) and machine learning engines, along with large language model platforms, have emerged over the last three decades. A notable early entrant, IBM's Deep Blue supercomputer made headlines in 1997 when it beat the world's reigning chess champion, Garry Kasparov. Similarly, the AlphaGo AI engine, developed by DeepMind, defeated the world's Go champion, Lee Sedol, in 2016. A watershed historical event at the time, it lost importance when a team of MIT researchers in 2023 was able to identify a strategic flaw. The team was able to beat AlphaGo with an engine called KataGo 100 percent of the time. However, team members had also discovered a weakness in KataGo that an amateur Go player, by the name of Kellin Pelrine, was able to mimic. He beat the KataGo AI engine fourteen out of fifteen games. As an interesting side note, he applied an initial circling strategy while feigning moves in another segment of the board as distractions. Pelrine indicated that the strategy does not work well on humans, as they notice it very quickly.

What Is the Historical Importance of the AI Gamesmanship?

The rapid advancement of AI engines has been marked by significant achievements, yet they still make mistakes. Nevertheless, it is inevitable that Total Experience and AI will become synonymous. Lightning-fast data processing is the strength of an AI engine, and it is the foundation for Total Experience performance assessment. In fact, you are already leveraging AI functionality for your business activities. If you search for something via Bing, Google, Safari, or any other search engine, you are leveraging AI. Similarly, if you bank with a large national or regional institution, your business benefits from AI for cybersecurity, transaction processing, and customer service activities. Additionally, your phone system's sharing of caller identification leverages AI. The intent of this observation is to alleviate fear surrounding AI and machine learning platforms. They are tools that your business is already enjoying benefits from using.

What Are the Next Steps in This Evolution?

News media has been abuzz about OpenAI and its AI product ChatGPT. When I queried ChatGPT about the history of OpenAI, I received this response:

> "OpenAI was founded in December 2015 by a group of technology leaders and entrepreneurs, including Elon Musk, Sam Altman, Greg Brockman, Ilya Sutskever, Wojciech Zaremba, and John Schulman. The initial aim of OpenAI was to advance research and development in

the field of artificial intelligence (AI) and ensure that the benefits of AI are distributed broadly for the betterment of humanity. The founders were motivated by the recognition that artificial general intelligence (AGI), or highly autonomous systems that outperform humans at most economically valuable work, could have profound societal impacts. To address potential risks and guide the development of AGI in a beneficial direction, OpenAI committed to principles such as broadly distributed benefits, long-term safety, technical leadership, and cooperative orientation."

The company started as a nonprofit entity to enhance the availability of machine learning and AI to the public, not just the large corporate entities with deep pockets. OpenAI created ChatGPT based on a large language model platform to allow users to interact with it using everyday speech—asking questions and receiving answers in various formats or alternative languages. In November of 2022, it launched the free version 3.0 of ChatGPT for public availability. This version was based on information available on the internet as of early 2021. The subscription-based ChatGPT 3.5 contained enhanced logic and had internet information as of January 2022. It became the 'free' version when the subscription-based version 4 became a revenue-generating product for the company in March 2023. Concurrently, OpenAI released application programming interface versions of their products. These applications enable ChatGPT integration with other software platforms (for example, Customer Relationship Management platforms, Customer Experience platforms, data analytics engines, or anything with a compatible application programming interface).

Not to be outdone, Google launched its Gemini AI tool (formerly known as Google Bard) in March 2023 as a free-to-use product. Besides cost, the key difference between Gemini and

ChatGPT-4 is that Gemini can access real-time internet data. ChatGPT is only knowledgeable as of January 2022 internet content. That said, the size of ChatGPT's model is 27 percent larger than Gemini. In theory, the larger model size means it can give better, more accurate answers when comparing data.

Similarly, Elon Musk's X.com (formally Twitter.com) released xAI Grok in November 2023. Upon launch, Grok had a long waiting list of eager users, as scaling the platform was a concern. Furthermore, you had to be a paying subscriber of X.com to access the platform. On the surface, Grok is more akin to Google Gemini, offering real-time access to the internet. However, it differentiates itself by injecting more humor and wit into its responses to questions. That creates a feeling of interacting with a real person versus an AI engine.

While most large language model AI engines do an excellent job of formatting text, providing sample letters for various tasks, and so forth, they do share a common challenge. As of this book's publication date, they often hallucinate their responses to direct information requests. When asking about specific people, events, or topics, they often blend the data elements, resulting in answers that may not be entirely factual. It is worth noting that the paid subscription version of ChatGPT is significantly better than its free predecessor, although it still suffers from occasional response hallucinations.

Why Is This Important to Business Leaders?

AI technology is firmly entrenched and will continue to evolve at an incredibly rapid pace. In the very near future, these tool sets will be as important to modern businesses as the telephone or the internet has been since the turn of the twenty-first century. Business owners and leaders who embrace AI and machine learning to enhance Total Experience will likely remain competitive. Those unable or unwilling to incorporate appropriate levels of engagement will risk falling behind their competition.

Need a Historical Example?

In the 1970s, personal computers were considered novelties and were prohibitively expensive by contemporary standards. During that era, most leaders could not imagine a personal computer as a valuable tool for their business. They had typewriters, fax machines, adding machines, and overhead projectors to complete their business operations. Fast forward to today, and if you were to ask business owners to choose between their personal computer or an office telephone, most would opt for the personal computer. This same sentiment will apply to AI in the next five to ten years. Business leaders will increasingly value no-cost/low-cost AI support over any other tool in their office. It is already available on their mobile devices. It is just a matter of time before it replaces much of the need for a physical computer or office desk telephone.

Should Business Owners Have Cybersecurity Concerns with AI Engines?

Absolutely! However, safeguarding critical business data in this evolving landscape will become increasingly difficult. The influence of large AI models will extend everywhere. In the short term, you absolutely must avoid exposing customer or employee personal identifiable information and sensitive financial details to these engines. All businesses should establish an office policy outlining the acceptable usage of AI, including ChatGPT, Gemini, Grok, or any other AI engines. You should train your employees in how to responsibly use the available AI tool sets. This will help minimize business risk while harnessing AI for efficiency gains in routine tasks. This proactive approach will optimize AI utilization and mitigate potential data loss and negative impacts on customers and employees.

Can You Leverage AI for Data Comparisons or Reporting and Still Maintain Cybersecurity?

Yes, by employing strategies to pseudo-anonymize your data (also referred to as *pseudonymization*). This will help to protect the identity of the individuals referenced in your data structure. For instance, when merging two datasets with a common data key, such as a client's first and last name, an AI model can easily process historical revenue and support costs in an Excel spreadsheet or Google Sheets format. To avoid providing raw information to an AI engine, assign a unique client identification (ID) number to each account. The files given to the AI application would then use the account ID, excluding client data. It is crucial never to share the map between the client name and the client ID, and to exclude datasets containing sensitive information like home address, phone number, birthdate, and the like.

In full disclosure, these anonymization methods will lose effectiveness as the AI engines evolve and grow. Assuming the same AI engine eventually will begin triaging your incoming phone calls, assisting with your email activity, helping with employee expense reports, and assisting with completing your tax returns, it will quickly know more about your business than you do. Still, there are moral and legal obligations to try to protect everyone's personal data.

Do Cloud-Based Total Experience Applications Leverage AI Processing?

With some degree of certainty, any vendors that provide a cloud-based offering—that use Amazon Web Services, Google Cloud, Microsoft Azure, or any of the other large providers—are currently harnessing AI functionality. While they may not be specifically analyzing the intricacies of your data, they are analyzing the digital traffic to identify potential cybersecurity threats.

Additionally, numerous Customer Experience and Customer Relationship Management software providers have enabled proprietary AI models into their applications. For example, Salesforce created Einstein, an AI engine focused on processing Customer Relationship Management activities associated with customer contacts, marketing campaigns, and more. Microsoft Dynamics 365 incorporates an AI model developed by Microsoft but now has enhancements offered by ChatGPT integration.

Do Business Owners Have Any Legal Relief with Respect to AI Use and Data Privacy?

As of the crafting of this book, the remedies offered in the United States are limited. Google, Microsoft, and OpenAI have data privacy policies that outline the data they collect, how long they will retain it, and their intended purposes. However, an AI engine does not store data in a traditional database format. So, a parent company's ability to monitor specific data retention is limited. From a United States regulatory perspective, the Computer Fraud and Abuse Act is the primary legal framework that might help govern the AI industry. Likewise, the Federal Trade Commission and the Securities and Exchange Commission might bring enforcement actions if AI companies fail to disclose cybersecurity risks or events to the public.

The terms of use and remedies for your data security within the various Customer Experience and Customer Relationship Management platforms vary significantly. When conducting due diligence on which products are best suited for your business needs, the terms of service and data privacy policies for each company should be important variables in the decision.

This contract review and remedy guidance should extend past the AI space. It is essential to review all software and service vendor contracts for cyber security liability. For instance, I recall reviewing a data privacy provision for a small vendor providing a software application. It supported employee recruiting and lead-generation activity. I was appalled to see that the terms of the vendor's service agreement removed all vendor liability for a data breach and placed full responsibility on the customer. To add insult to injury, according to the terms in the service contract, the customer was liable for the costs that the vendor might incur in trying to cure the security breach (which the vendor's cybersecurity negligence may have caused). It was an unreasonable and crazy contract. However, according to the vendor's sales rep, the company had never been challenged to change it.

CHAPTER 11 SUMMARY

- Artificial intelligence (AI) is becoming increasingly important and will drive Total Experience for virtually all business owners and leaders in the future.
- AI is already being used to support small business activity in several industries, such as search engines, banking, and telecommunications.

- General public access to relatively accurate AI responses became readily available in 2023 with the introduction of ChatGPT, Gemini, and Grok.
- Cybersecurity concerns remain with AI use; caution should be taken before giving AI models protected private customer or employee information.
- Various Total Experience vendors are leveraging AI processing within their applications today—business owners and leaders should understand vendor data protection policies when choosing a product for their business.

CHAPTER 12

The Future of Total Experience

Total Experience is poised to expand its focus for businesses of all sizes. This book has delved into several areas that influence Total Experience for customers, employees, and other stakeholders of your business. This insight is derived from current events, historical activity, and future predictions. One subsequent prediction is that Total Experience will expand into the realm of forecasted behavior. Advancements in AI and cyber technology, coupled with exposure to new data structures, are set to make the forecasting of business events very accurate.

What Are Some Examples of This Evolving Landscape?

Consider the ongoing initiative in the pet care industry to develop wearable devices to monitor the physical activity of dogs and cats. This effort aims to track calorie burn and alert pet owners to physical issues or behavior concerns. This technology would allow pet day-care businesses to have another indication of whether their attendants are adequately exercising the animals. Likewise, veterinarians could monitor a pet's well-being after the animal returns home from a surgical procedure. With this data and the power of AI engines, predicting animal health becomes possible.

Picture a pet food or supplement supplier armed with this information from a hundred million pets. Similarly, consider the value of this information for an insurance company offering medical policies for insured animals. Insurance company cost savings could be massive.

We touched on this forecasting concept in Chapter 4 during our in-depth exploration of Workforce Management. It highlighted how these platforms utilize historical contact information to help forecast future contact volumes and the associated staffing needs. Current Workforce Management models employ fairly basic mathematical algorithms. When compared to those utilized by AI engines like ChatGPT, it is like comparing a car from the 1950s to a current Formula One race car. Most schedule management applications in use today operate with static logic. This constrains their ability to accurately forecast future events. In contrast, AI engines consistently learn and adapt with each passing day.

This decade offers a new host of technology capabilities with the introduction of AI in the form of large language models. These tools are now available to businesses of all sizes for minimal cost. This is a complete economic game changer. Consider this: just seven years ago, utilizing IBM Watson to automate customer service activities via chatbot models required an exceptionally large checkbook. Moreover, it required a significant amount of engineering time to fine-tune Watson's responses to questions.

Figure 12.1 depicts a model of the evolution of data management. The speed of advancement in new capabilities is increasing.

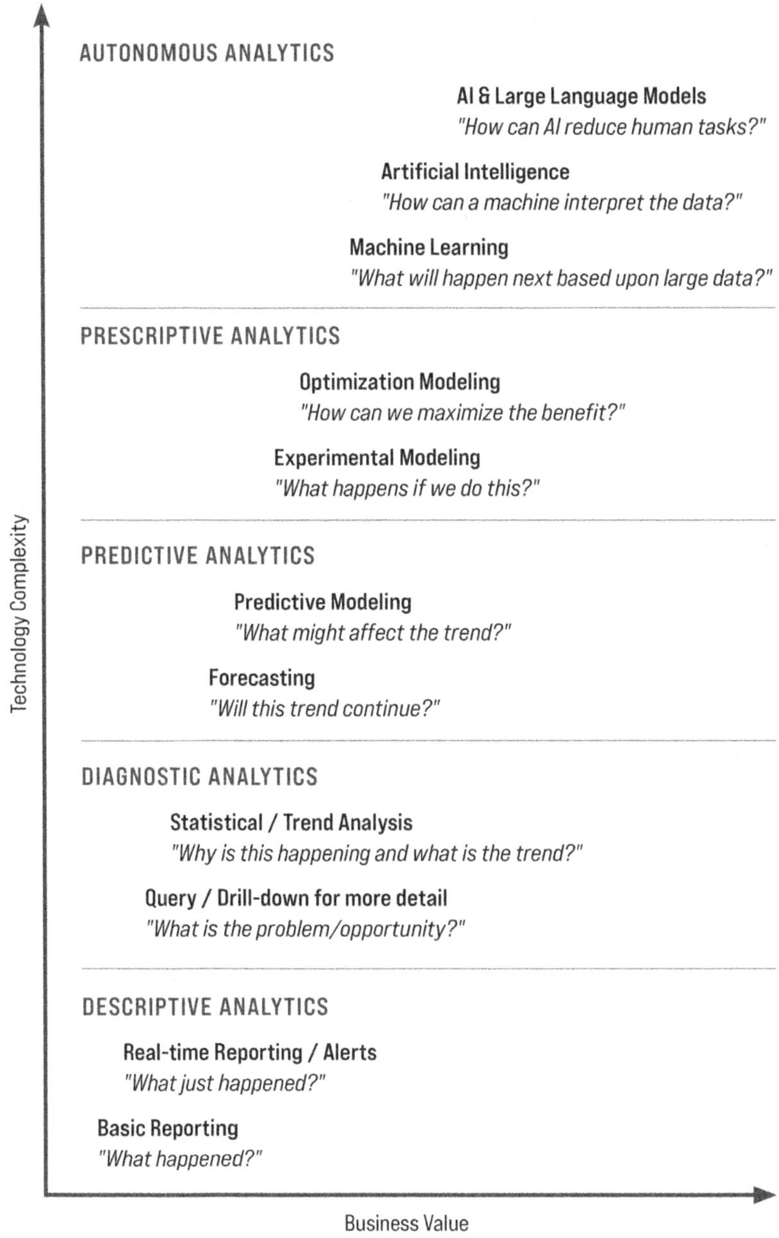

AUTONOMOUS ANALYTICS

AI & Large Language Models
"How can AI reduce human tasks?"

Artificial Intelligence
"How can a machine interpret the data?"

Machine Learning
"What will happen next based upon large data?"

PRESCRIPTIVE ANALYTICS

Optimization Modeling
"How can we maximize the benefit?"

Experimental Modeling
"What happens if we do this?"

PREDICTIVE ANALYTICS

Predictive Modeling
"What might affect the trend?"

Forecasting
"Will this trend continue?"

DIAGNOSTIC ANALYTICS

Statistical / Trend Analysis
"Why is this happening and what is the trend?"

Query / Drill-down for more detail
"What is the problem/opportunity?"

DESCRIPTIVE ANALYTICS

Real-time Reporting / Alerts
"What just happened?"

Basic Reporting
"What happened?"

(Y-axis: Technology Complexity; X-axis: Business Value)

FIGURE 12.1 A model of the evolution of data management and analytics

To illustrate, a large bank in the Asia-Pacific region invested six months and millions of dollars to train the platform to handle one complex customer service workflow. Company engineers and IBM representatives were rightly quite proud of their achievement. However, given the multitude of workflows within the bank, any hope of completely replacing human customer service agents with Watson was going to be a long journey. Fast forward to today: companies can automate many similar workflows in Watson or other AI platforms in a matter of weeks or even days.

By projecting forward just five years into the future, the potential capabilities are staggering. Envision having access to an AI engine that comprehends all your business activity. This includes your clients, the services or products you provide, sales frequency, and every aspect of your operations. Now, consider that same AI engine tapping into information about everyone in your community, state, or across the country. This wealth of data, coupled with bank transaction history, telecommunication events, purchase activity, tax information, and more renders an AI engine immensely powerful. The ability to forecast who will require your services, predict when they will reach out, assess their creditworthiness, and recommend the optimal portfolio of services or products for a successful sale is not out of the question.

Scary? For many, the answer is "yes." However, we already have a taste of this. Who has not performed a web search and subsequently found themselves bombarded by advertisements? Even after clearing your web browser history or switching devices, the onslaught of ads persists.

Now imagine owning a business without this level of market visibility and trying to compete with another business that has it. You might persist for a while with loyal customers who appreciate your traditional business practices. However, over time, they will be replaced by customers who expect Total Experience engagements to be on par with their other vendors.

This business evolution will also extend to employee expectations. Returning to Maggie and her nursing business, many nurses will become accustomed to querying an AI engine for information about their work schedule and clients. Envision nurses being able to interact with the same AI engine while traveling to a new client hospital for their initial shifts. They could verbally inquire about the current patients they will be assisting. The AI engine will help them with the route to the hospital and provide parking information. Naturally, the software application, having already made the necessary hotel reservations, would provide a list of restaurants in the area. Those recommendations will consider unique food preferences and allergies. The nurses will not need to track trip expenses. Rather, the AI engine will have all those events captured. And when they complete their trips, it will automatically submit the expense reports, saving the employees an hour of personal time. No accounts payable clerk would need to audit the report, as the AI engine would process it during the next payroll run. Lastly, the engine will monitor road conditions in advance of the nurses' return trips. It will automatically provide instructions for the best routes home. In the event of issues during travel, it would automatically prepare lodging and communicate delays to relevant parties.

So, here is the question: If you own a competing nursing service but do not have a comparable Total Experience, how hard would it be to recruit new nurses or retain existing ones? Even small businesses will need to implement AI-supported Total Experience workflows to stay competitive in their labor market. A hesitant owner may attempt to compensate employees for lack of AI services via increased pay rates. However, trying to stay in business with that added headwind may prove impossible.

Extending that scenario further, I envision Maggie navigating a changing regulatory environment and trying to stay abreast of new federal and state regulations. She could opt to hire a legal service to assist with compliance efforts or leverage a legally trained AI engine to monitor the regulatory landscape and notify her of any changes. This same engine could also help construct and submit the necessary reports for business compliance. This could save significant staff administration labor hours versus trying to manually construct the reports and submit them. Impossible? Not at all. ChatGPT-4 completed the Uniform Bar Examination and achieved a passing score in 2023.

What Near-Term Technologies, Other than AI, Will Influence Total Experience?

While no one has a crystal ball to accurately predict the future, blockchain and mixed reality are two technologies that are poised to disrupt current business processes.

Blockchain, widely used in the cryptocurrency industry and by many banking institutions, creates a distributed transaction ledger on the internet. This ledger tracks all aspects of a specific financial event, stores it on numerous computing platforms, and ensures cybersecurity by preventing the falsification of one record without affecting all related records. The software industries affecting Total Experience are already starting to adopt this distributed ledger concept to track customer interactions, manage supply chain events, and store digital identities online, which enhances cybersecurity measures.

Will Blockchain Expertise Be a Requirement for Business Owners and Leaders?

No. The Total Experience software applications will leverage blockchain in the same manner as they use different programming languages to perform different logical tasks. Today, you do not have to be a Python developer to add customer contact information into a Customer Relationship Management system. In that same manner, tomorrow you will not need to become a blockchain expert to complete workflow processes leveraging the technology. Additionally, with the help of your integrated software applications, you will be able to easily dissect any blockchain transaction for reporting.

What about Mixed Reality?

Mixed reality combines physical world events with the digital world. While it has been prevalent in the gaming industry for a few years, it has yet to impact general business applications. However, I anticipate that status is about to change.

Before the COVID-19 pandemic, I had the opportunity to attend a conference and get a VIP review of Magic Leap, the virtual reality headset that is trying to lean into business settings with mixed reality technology. My first amazement with this company's technology was seeing a *Star Wars*-like battle game being played with characters on a piece of furniture in the room. Specifically, the animated game pieces battled each other in 3D life-like form. The game pieces were visible via the headset that I wore. However, a cool feature was that I could place them on a random piece of furniture (like a coffee table), and the characters were cognizant of the furniture's edge. Moreover, they were cognizant of the perils of falling from it. They became injured or died when falling from the table and hitting the floor.

Fun Stuff, but How Does It Apply to a Business Setting?

I observed two real-world scenarios where technology could be helpful. The first was dealing with a virtual, 3D portrayal of a 1950s pickup. You could select assemblies like the engine and pull it apart to see the inner workings and construction. Other conference attendees were in the room with similar Magic Leap headsets. We could synchronize our headsets to share the viewing of the 3D model. We each had a handheld controller that allowed us to select an item from the vehicle, like the transmission, pull it apart and explode the internal view to see the other components. We could effectively team up to tear down the pickup's assemblies or reconstruct them.

The best movie analogy I can offer is one of the Tony Stark (Ironman) invention scenes where he explodes the architecture of his father's campus, strips away noise, and sees the architecture of a new power supply. Very cool!

The second scenario was a 3D virtual medical perspective of the human brain with a tumor. The digital representation was from a series of MRI images. Participants wearing the headsets could use the associated hand controllers to magnify the brain, the tumor, and all associated body parts in a 3D manner. You could change the perspective 360 degrees or move into underlying layers of tissue to see capillaries and neurosurgical components. The training scenario was aligned with the use of various medical instruments to cut away the tumor. Thereby, the baseline assumption was that the MRI and Magic Leap would allow the surgical team to gain experience in a presurgical training exercise before conducting the actual surgery.

Any Ancillary Concerns Related to Total Experience and the Evolving Technology?

I do have a few worries regarding the increasingly fast development of various innovative technologies. AI development is at the top of the list. Therein, I hope business leaders will still have a sense of humanity in their approach to Total Experience. It will be incredibly sad if leaders adopt the unfeeling, nonhuman advice of an AI engine for employee interactions or community events like natural disasters. I have doubts that new technology offerings will have the same sense of compassion that we humans bring to everyday business transactions.

Further, I can envision a landscape where corporate-sponsored AI engines strive to influence one another. In theory, this could be completed like a cyber-denial-of-service attack by one AI flooding the other model with incorrect or irrelevant content via source data channels. Still more nefarious, the machines could try to disable each other's infrastructure by sabotage. Hopefully, we humans and our businesses are not collateral damage in those scenarios. Based on recent media reports, it seems Elon Musk shares concerns for the uncontrollable evolution of technology. Pandora's box has been opened. We humans will have to adapt and try to manage within its limits.

Those worries aside, I see a bright future with AI, blockchain, mixed reality, and other technological advancements, full of potential for companies big and small. The efficiencies they bring to our day-to-day activities will enhance Total Experience for our clients, employees, and other stakeholders. In theory, administrative costs should decline. Therefore, all businesses should benefit from the elimination of repetitive tasks and avoidable rework.

A Wish and a Prayer

As the world speeds toward technology enablement and more transactional relationships, the focus on good ethics seems to be waning. Companies, universities, and governments often prioritize short-term optics over fostering a culture of sustained trust, responsibility, and meritocracy. We are witnessing new quality issues in industries like aircraft manufacturing and transportation management that seem to be ethics-based and that were largely absent in past decades. Instances of plagiarism and cheating among university staff members and students are becoming frequent. Sadly, they are being met with limited consequences. Governments around the world are being found guilty of promoting corruption and graft. Many of today's challenges stem from a decline in ethical behavior across the globe.

It is not surprising that I believe maximized Total Experience is impossible without a culture grounded in business ethics. This environment starts with the leadership team, extends to all employees, and should be table stakes for any affiliated business relationships.

Examples of ethical business behavior can still be found in rural areas around the globe. Business leaders often approach their employees as family members and strive for fairness. Customer relationships are managed ethically with a focus on caring for neighbors rather than pursuing "one and done" interactions. Moreover, service providers and vendors are typically local businesses, owned by neighbors; business owners and leaders give them ethical consideration and enjoy long-term relationships that strive for fairness.

My wish is that business owners and leaders who read this book will apply ethical principles within their own organizations and serve as examples for others. Additionally, it would be excellent to have school administrators, at all levels, incorporate the study of ethics into their curricula. The prayer is that the combination sparks a culture movement that has a global, Total Experience positive impact.

Does Total Experience Really Matter?

In the end, Total Experience methodology is irrelevant unless it helps your business thrive and extends its long-term viability. Other philosophies and methodologies may be a better fit for your specific organization. That said, I cannot envision any successful business models that do not address the voice of your customers, the voice of your employees, the needs of your strategic partnerships, and the overall efficiencies of your business operations.

In summary, I passionately believe those business owners and leaders who find a balance between Total Experience investment and business benefit will thrive. For everyone else, the Will Rogers quote still applies: "Even if you are on the right track, you'll eventually get run over if you just sit there."

CHAPTER 12 SUMMARY

- Artificial intelligence (AI) will be a significant driver of Total Experience in the future.
- Blockchain will play a crucial role in the future of Total Experience by combating fraud and enhancing cybersecurity.
- Mixed reality will be increasingly used for Total Experience activities supporting training, activity preparation, and testing.

- Companies that prioritize Total Experience methods and cost minimization are more likely to survive than those that do not.
- Do not underestimate the power of good ethics in leadership and business success.
- Our future is what we make of it—choose wisely.

GLOSSARY

Account Receivables (A/R): The total amount of money owed to a company by its customers.

Add-on: An additional feature or service that can be purchased with a software application.

AI (artificial intelligence): A branch of computer science concerned with creating intelligent machines that can perform tasks typically requiring human intelligence.

AI engine: A computer platform that can learn on its own by processing new data.

AI machine learning model: A type of AI engine that is trained on a large amount of data to make predictions about future events.

Amazon Web Services (AWS): A cloud computing platform offered by Amazon.

Application integration: The process of connecting two or more applications, so they can share data and functionality.

Application Programming Interface (API): A set of routines, protocols, and tools to build software applications.

Average call handing time: The average amount of time it takes a customer service agent to handle a call.

Average labor rate: The average hourly wage paid to employees.

Average number of calls per period: The average number of calls that are received by a customer service agent in each period (for example, hour, day, week).

Bill of materials (BOM): A list of the parts and materials needed to make a product.

Blockchain: A distributed ledger technology that allows for secure and transparent transactions.

Brand: A multifaceted concept in business that constitutes an entity's identity, image, reputation, value proposition, and customer emotion.

Brand loyalty: The customers' commitment to a particular brand and their willingness to repurchase its products or services.

Business ethics: The moral principles and values that guide individuals and organizations in business by leveraging integrity, fairness, honesty, and respect for all parties being interacted with.

Call routing: The process of directing incoming calls to the appropriate agent or department.

Cloud based: Software or services that are delivered over the internet and can be accessed from any device.

Computer Fraud and Abuse Act (CFAA): A law in the United States that prohibits unauthorized access to computers and computer systems.

Concurrent seat licensing: A licensing model where the number of licenses is based on the maximum number of users that will be using the software at the same time.

Contact center: A department or facility that is responsible for managing customer interactions.

Contact Center as a Service (CCaaS): A cloud-based software service that provides contact center functionality, including call routing, Interactive Voice Response, and agent management to businesses.

Contact channel: The method used by a customer to contact a business, such as phone, email, or online chat.

Continuous improvement: A continuous process of making minor changes to improve a product or service.

Cost of Operational Support: The cost of supporting the ongoing operations of a business, including customer service, marketing, and administration.

Cost of Product Delivery: The cost of designing, producing, and delivering a product or service to the customer.

Customer Experience (CX): An industry focused on enhancing the customer's perception of all interactions with a business, from awareness to purchase and beyond.

Customer Relationship Management (CRM): A software application designed to help businesses manage their relationships with customers.

Customer satisfaction (CSat): The customer's perception of the quality of a product or service.

Cybersecurity: The protection of computer systems and networks from unauthorized access, use, disclosure, disruption, modification, or destruction.

Data analysis: The process of examining and interpreting data to gain insights.

Data center: A facility that houses computer systems and related equipment.

Data privacy: The protection of personal data from unauthorized access, use, disclosure, alteration, or destruction.

Data structures: The ways in which data is organized and stored.

Data visualization: The process of transforming data into a visual format, such as a chart or graph.

Day-to-day efficiencies: The ways in which a company can improve its efficiency on routine tasks.

Digital payments: Electronic payments made through the internet or other digital channels.

Effective interest rate: The actual interest rate paid on a loan, considering all fees and charges.

Employee churn: Slang for the rate at which employees leave a company.

Employee Experience (EX): The employees' perception of their work environment and interactions with the company.

Employee sentiment: The overall morale and satisfaction of a company's employees.

Enterprise grade: Software applications that are designed for large businesses.

Enterprise Resource Planning (ERP) system: A comprehensive software application that integrates and automates core business processes across an entire organization.

Excel: A spreadsheet application offered by Microsoft available as a desktop application or cloud based.

Federal Trade Commission (FTC): An independent agency of the United States government responsible for protecting consumers from unfair or deceptive business practices.

Fly-wheel effect: A cycle of events that reinforces itself, leading to more of the same.

Franchise fee: A fee that a franchisee pays to a franchisor.

Franchisor: A company that grants licenses to franchisees to operate its business model and use its brand.

Fraud event: An event in which someone attempts to deceive a business for financial gain.

Functional requirement: A requirement that specifies what a software application must do.

Google Drive: A cloud-based file storage service offered by Google.

Google Gemini: A large language model chatbot developed by Google.

Google Looker Studio: A data visualization platform offered by Google.

Google Sheets: A spreadsheet application offered by Google.

Google Workspace Business Plus: A suite of software applications offered by Google that includes Gmail, Google Drive, Google Calendar, and Google Docs.

Grok: A large language model chatbot developed by xAI.

Headwinds: Challenges or difficulties that a business is facing.

IBM Watson: A cognitive computing platform developed by IBM.

Ideal customer journey: A map of the steps that a customer takes from the first time they hear about a business to the time they become a loyal customer.

Identity and Access Management (IAM) software: Software used to manage user access to applications and data.

Implementation process: The process of installing and configuring a software application.

Inbound call: A phone call that is received by a business.

Insight data: Data that provides valuable information about your business.

Interactive Voice Response (IVR): An automated phone system that allows callers to interact with a business without speaking to a live agent.

Key Performance Indicator (KPI): A metric that is used to measure the performance of a business.

Large Language Model (LLM): A statistical method used to process human language.

Lean manufacturing: A manufacturing philosophy that focuses on eliminating waste.

Legal and Compliance: A section of a Request for Proposal that asks questions about the legal and compliance requirements of the software application.

Live agent: A customer service representative who interacts with customers in real time.

Lot control: The process of tracking and controlling the production of a product.

MRI (Magnetic Resonance Imaging): A medical imaging technique that uses a strong magnetic field and radio waves to create detailed images of the organs and tissues inside a living body.

Market share: The percentage of the total market that a company holds.

Master Service Agreement (MSA): A contract that establishes the basic terms and conditions for the provision of services.

Microsoft Azure: A cloud computing platform offered by Microsoft.

Microsoft Business 365 Premium: A suite of software applications offered by Microsoft that includes Exchange, OneDrive, SharePoint, and Office.

Microsoft Dynamics 365 Customer Engagement: A Customer Relationship Management platform offered by Microsoft.

Microsoft Power BI: A data visualization platform offered by Microsoft.

Mixed reality (MR): A technology that blends the physical world with the digital world.

Module: A separate component of a software application that provides specific functionality.

Named user license: A license that is assigned to a specific user and cannot be used by anyone else.

Net income: A company's total profit after all expenses have been paid.

Net Promoter Score (NPS): A metric that measures customer loyalty.

Number of emails worked per period: The number of emails that are handled by a customer service agent in each period (for example, hour, day, week).

Office administration: The administrative tasks that are necessary to run a business.

OneDrive: A cloud-based file storage service offered by Microsoft.

One-off sales: Sales that are not expected to be repeated.

OpenAI: A research laboratory focused on promoting and developing friendly artificial general intelligence.

Operational Excellence (OpEx): A business philosophy that focuses on continuous improvement, efficiency, cost reduction, and customer satisfaction.

Org Chart: Business slang for "organization chart" which depicts the reporting relationship for a business or non-profit entity.

Outbound call: A phone call that is made by a business.

Outsourced billing service: A service that handles billing for businesses.

Pain point: A problem or obstacle that makes it difficult for a business to achieve its goals.

Payroll: The process of calculating and paying employee wages.

Per unit of activity: A pricing model in which the cost of the software is based on the number of times it is used.

Personal Identifiable Information (PII): Information that can be used to identify an individual, such as their name, address, or Social Security number.

Pilot: An initial test of a new product or service.

Point-of-Sale (POS) system: A software application that helps businesses track sales and inventory.

Predictive power: The ability to predict future events with accuracy.

Pricing model: The way in which a product or service is priced.

Private Branch Exchange (PBX): A telephone system used by businesses to manage internal and external calls.

Process bugs: Errors or problems in a business process.

Process consultant: A professional who helps businesses improve their processes.

Process improvement: The process of making changes to a process to make it more efficient or effective.

Product bugs: Errors or problems in a product.

Product development: The process of creating new products.

Proof-of-Concept (POC): A demonstration of the feasibility of a new software application or service.

Published list price: The price of a software application that is listed in a vendor's catalog or website.

Purchase-to-delivery lead time: The amount of time it takes for a company to receive a shipment of goods after it has been ordered.

Raw material inventory: The quantity of raw materials that a company has on hand.

Recruiting process: The process of finding and hiring new employees.

Reporting or dashboarding segment: A section of a software application that allows users to view reports and dashboards.

Request for Proposal (RFP): A document that requests proposals from vendors for a specific product or service.

Revenue: A company's total income from sales.

SaaS application suite: A suite of software applications that are delivered over the internet and can be accessed from any device.

Salesforce.com: A customer management application platform supporting business sales, operations, and support.

Schedule models: Templates that can be used to create employee schedules.

Securities and Exchange Commission (SEC): An independent agency of the United States government responsible for protecting investors and maintaining fair, orderly, and efficient markets.

Segmentation: The act of dividing a group of people or things into smaller groups.

Selection process: The process that will be used to select a vendor.

Self-service: The ability of customers to help themselves without having to contact a customer service representative.

Service culture: A business culture that is focused on providing excellent customer service.

Service pricing: The price that a company charges for its services.

Sourcing decisions: Decisions about where and how to find new employees, software applications, products, or raw materials.

Stakeholder: A person or group of people who have an interest in a business.

Statement of Work (SOW): A document that outlines the specific work that will be performed by a vendor.

Tenured staff: Employees who have worked for a company for a long time.

Total Experience: A concept introduced by Gartner in 2020 that encompasses all aspects of a business product or service delivery, including Customer Experience, Employee Experience (EX), Multi-User Experience, and Operational Excellence. It emphasizes the need to consider all stakeholders and their interactions with the business to ensure the business is viable and thrives well into the future.

Total Quality Management (TQM): A management approach that focuses on continuous improvement.

Transaction fee: A fee that is charged for each transaction that occurs.

Virtual reality (VR): A technology that creates a simulated environment that you can interact with.

Visualization platform: A software application that allows you to create visualizations of your data.

Waiting list: A list of customers who are waiting for a product or service.

Weighting and scoring method: A method of evaluating proposals in which each criterion is assigned a weight, and each proposal is assigned a score.

White-glove service: A high-quality service that is personalized and tailored to the specific needs of the customer.

Work schedules: The hours that employees are scheduled to work.

Workflow applications: Software used to automate and manage business processes, such as customer service and order fulfillment.

Workforce Management (WFM) application: A software application that helps businesses manage their workforce, including scheduling, forecasting, and reporting.

Workforce planning: The process of planning and managing employees' schedules.

Y2K: Acronym for the Year 2000—when all the world's computer systems were supposed to fail due to the lack of databases being able to support the data schema of the new century (also known as the Millennium Bug).

Year-over-year revenue growth: The percentage change in a company's revenue from one year to the next.

BIBLIOGRAPHY

The following books and resources are preeminent materials for deeper topical research:

Total Experience

- *Gartner Top Strategic Technology Trends for 2021* by Kasey Panetta (October 19, 2020)

Customer Experience

- *Service Operations Management, Second Edition: The Total Experience* by David W. Parker (2018)
- *The Service Culture Handbook: A Step-by-Step Guide to Getting Your Employees Obsessed with Customer Service* by Jeff Toister (2017)

Operational Excellence

- *Lean Production Simplified: A Plain Language Guide to the World's Most Powerful Production System*, 3rd Edition by Pascal Dennis (2015)
- *The Goal: A Process of Ongoing Improvement* by Eliyahu M. Goldratt and Jeff Cox (1994)

Data Analytics

- *Competing on Analytics: The New Science of Winning* by Thomas H. Davenport and Jeanne G. Harris (2007)
- *Lean Analytics: Use Data to Build a Better Startup Faster* by Alistair Croll and Benjamin Yoskovitz (2013)

Business Support

- *How to Downsize Your Business Successfully (Entrepreneur Today Series Book 8)* by Mark Patene (2014-15)

Artificial Intelligence

- *Adversarial Policies Beat Superhuman Go AIs* by Tony T. Wang, Adam Gleave, Tom Tseng, Kellin Pelrine, Nora Belrose, Joseph Miller, Michael D. Dennis, Yawen Duan, Viktor Pogrebniak, Sergey Levine, Stuart Russell *(November 1, 2022)*

INDEX

Note: Page references in italics indicate figures.

account receivables (A/R), 115
ACH (automatic clearing house) transactions, 37
add-ons, 115
ADP Payroll, 28
Adyen, 22
AI. *See* artificial intelligence
Alchemer, 23
AlphaGo, 93
Altman, Sam, 94
Amazon, 19
Amazon Connect, 19
Amazon Web Services (AWS), 99, 115
answering services, 20
Apple, 82
Apple Pay, 22
application integration, 115
application programming interface (API), 95, 115
artificial intelligence (AI), xiii
  AI engines, defined, 115
  and anonymization, 98
  ChatGPT, 94–99, 104, 108
  with cloud-based applications, 99
  and cybersecurity, 97–98
  defined, 115
  emergence of, 93
  Gemini, 95–96, 118
  Grok, 96, 119
  historical importance of AI gamesmanship, 94
  importance to business leaders, 96–97
  large language model engines, 95–96, 104
  and legal relief with respect to data privacy, 99–100
  machine learning model, 115
  OpenAI, 94–95, 99, 121
  summary of, 100–101
Aspect, 13, 18, 30
AT&T, xi, 19, 27–29
automation, 12–13, 59, 74
Avaya, 13, 18, 30
average call handing time, 29, 115
average labor rate, *85*, 85–86, 115
average number of calls per period, 29, 115

bankruptcies, ix
Berkshire Hathaway Inc., 88
Bill Me Later, Inc., 88

127

bill of materials (BOM), 46, 116
Block, 22
blockchain, 108–109, 116
Bloomerang, 22
brand
    awareness of, *69*
    defined, 116
    loyalty to, 6, 81
    use by franchisors, 118
Brockman, Greg, 94
business consultants, 61, 70
business ethics, 112, 116

calls
    average handing time, 29, 115
    average number per period, 29, 115
    inbound, 119 (*see also* Interactive Voice Response systems)
    outbound, 17, 121
    routing of, 112, 116
casework assignments, 17
cash flow, 68, *69*
CCaaS. *See* Contact Center as a Service
Ceridian, 28
CFPB (Consumer Financial Protection Bureau), 38–39
Chapel, Maggie (case study)
    nursing business of, 2–4, 13, 30, 59, 73–74, 107–108
    nursing-career dream of, 1
    on recruiting nurses, 1–2

charging stations, 7
chatbots, 23
ChatGPT, 94–99, 104, 108
Cisco, 13, 30
cloud-based software or services
    AI with, 99
    Contact Center as a Service, 14, 18–19, 116
    defined, 116
    Google Drive, 118–119
    growth of, 21
    licenses for, 18–19
    OneDrive, 31, 120–121
Computer Fraud and Abuse Act (CFAA), 99, 116
concurrent seat licensing, 18, 116
Consumer Financial Protection Bureau (CFPB), 38–39
Contact Center as a Service (CCaaS), xi, 14, 18–19, 116
contact centers, 18–20, 32, 34, 116
contact channels, 17, 30–33, 116
continuous improvement, 4–5, 116
cost of operational support, 117
cost of product delivery, 6, 117
COVID-19 pandemic, 67
CRM. *See* Customer Relationship Management
CSat. *See* customer satisfaction
Customer Experience (CX), xii. *See also* Contact Center as a Service

Index | 129

achieving good technology value, 25–26
at Amazon, 19
application features, variety in, 24–25, *25*
and casework assignments, 17
contact channels' use in, 17
Customer Relationship Management systems for, 21–22
customer satisfaction surveys, 23–24
defined, 117
at eBay, 10–12
growth of, 17
Interactive Voice Response systems' role in, 14
McCabe on, 6
outsourcing of services, 20
and pricing models, 19–20
and reporting software, 17–18
Salesforce, 21–22
self-service, 22–23
software platforms for, 20–21
summary of, 26
and technology licensing, 18–19
Customer Relationship Management (CRM), 21–22, 117
customer satisfaction (CSat), 117
cybersecurity
and AI, 97–98
at AT&T, 27–28
and blockchain, 108
defined, 117

data analysis, 117
data centers, 27, 41, 117
data privacy, 99–100, 117. *See also* cybersecurity
data structures, 103, 117
data visualization, 117–120
day-to-day efficiencies, 75, 117
Deep Blue supercomputer, 93
DeepMind, 93
Deming, W. Edwards: *Quality, Productivity and Competitive Position*, 59
digital payments, 22, 117
Discover, 37
Dodd-Frank Wall Street Reform and Consumer Protection Act, 38
Donahoe, John, 10–12
DonorPerfect, 22
Drucker, Peter, 13

eBay, x–xi, 9–13, 37–38, 69
effective interest rate, 83, 117
Einstein (AI engine), 99
electric vehicles (EVs), 6–7
employee churn, 85, 118
Employee Experience (EX), xii
at AT&T, 27–29
contact channels for, 30–33
defined, 118
employee satisfaction targets, 29
and employee surveys, 29
and feedback sessions, 29

and Human Capital
Management systems, 28
and information technology
support, 29
and labor cost
management, 34
McCabe on, 6
payroll and benefits, 28
and spreadsheets
for workforce
management, 30–31
summary of, 34–35
and workflows, 29
and Workforce Management
applications, 29–34
employee sentiment, 67–68, 118
enterprise-grade
applications, 30, 118
Enterprise Resource Planning
(ERP) system, 118
EVs (electric vehicles), 6–7
EX. *See* Employee Experience
Excel, 31–33, 53, 98, 118

Federal Trade Commission (FTC),
99–100, 118
fly-wheel effect, 86, 118
Forrester Research, Inc., 3
Fortune 500 companies, 14
franchise fees, 61, 118
franchisors, 61, 64, 70, 118
fraud events, 17, 118
FTC (Federal Trade Commission),
99–100, 118
functional requirements, 118

future of Total Experience, xiii–xiv
and blockchain, 108–109
and business ethics, 112–113
competitive
advantage, 106–108
concerns about, 111
data management and
analytics, 104, *105*, 106
examples of the evolving
landscape, 103–104,
105, 106–108
medical applications, 110
mixed reality, 108–
111, 113, 120
near-term technologies, 108
pet monitors, 103–104
summary of, 113–114
virtual reality, 109–110

Gartner Inc., 3, 6, 123
*Gartner Top Strategic Technology
Trends for 2021* (Panetta), x
Gateway Inc., x, 45–47
General Electric, 49
Genesys, xi
getting started in Total
Experience, xiii
automation, 59
best practices before
launching a business,
establishing, 64
business
consultants, using, 61
data collection, 60

Index | 131

employee support for solutions, 62–63
issue resolution options, 60–61
Maggie's nursing business (case study), 59
pain points, 58–64
piloting solutions, 61–63
reassessing success against goals, 63–64
stakeholders, 59, 63
steps (series one), 58
steps (series two), 58–59
summary of, 65
and survival rate for new franchises, 64–65
Givebutter, 22
Gmail, 119
Google Calendar, 119
Google Cloud, 99
Google Docs, 119
Google Drive, 31, 118–119
Google Gemini, 95–96, 118
Google Looker Studio, 52–53, 118
Google Sheets, 31, 53, 98, 119
Google Workspace Business Plus, 31, 119
Grok, 96, 119

headwinds, 67, 119
Hoag, Tom, 76–77
Home Instead, x–xi
How to Downsize Your Business Successfully (Patene), 79

Human Capital Management systems for Employee Experience, 28

IAM (Identity and Access Management) software, 12–13, 119
IBM, 106
IBM Deep Blue, 93
IBM Watson, 104, 106, 119
ideal customer journey, 65, 119
Identity and Access Management (IAM) software, 12–13, 119
IDEX Corporation, x–xi, 76–77, 88–89
implementation process, 119
inbound calls, 119. *See also* Interactive Voice Response systems
insight data, 119
Interactive Voice Response (IVR) systems, 13–14, 23, 59, 116, 119
International Franchise Association, 64–65
iPhone, 82
IVR (Interactive Voice Response) systems, 13–14, 23, 59, 116, 119

JD Edwards World, 46

Kasparov, Garry, 93
KataGo, 93

Key Performance Indicators (KPI) (Marr), 70
Key Performance Indicators (KPIs), xiii, 67–71, *69*, 86–87, 91, 119

large language model (LLM), 93, 95–96, 104, 118–119
lean manufacturing, 81, 119
legal and compliance requirements, 119
live agents, 23, 119–120
LLM (large language model), 93, 95–96, 118–119
lot control, 120

Magic Leap, 109–110
magnetic resonance imaging (MRI), 110, 120
market share, 68, *69*, 120
Marr, Bernard: *Key Performance Indicators (KPI)*, 70
Mastercard, 37
Master Service Agreement (MSA), 45, 120
McCabe, John, 6
Medallia, 23
mergers and acquisitions, 88–90
Microsoft Azure, 99, 120
Microsoft Business 365 Premium, 31, 120
Microsoft Dynamics 365 Customer Engagement, 22, 99, 120

Microsoft Power BI, 52–53, 120
mistakes to avoid, xiii
  being oversold on service offerings, 82–83
  committee-based KPI accountability, 86–87
  disruptive mergers and acquisitions, 88–90
  farmers' avoidance of mistakes, 83–84
  low wages leading to high employee churn, 84–86, *85*
  over-focusing on specific enhancement programs, 81–82
  relying on outdated data, 87–88
  selecting a solution before you understand the problem, 90–91
  summary of, 91
  trying to fix everything at the same time, 60
MIT, 93
mixed reality (MR), 108–111, 113, 120
modules, 33, *43*, 120
Motorola, 81–82
MR (mixed reality), 108–111, 113, 120
MRI (magnetic resonance imaging), 110, 120
MSA (Master Service Agreement), 45, 120
Multi-User Experience, xii, 4

with PayPal, 38–39
proposal for service
for, 44–45
scope of, 39
for small businesses, 44–45
software/service applications
dedicated to, 40 (*see also*
Request for Proposal)
summary of, 47–48
Musk, Elon, 6–7, 94, 96, 111

named user licenses, 18,
41, *43*, 120
natural language, 14
net income, 68, 120
Net Promoter Score (NPS),
10–11, 69, 120
Nokia, 82
nonprofit donation
management, 22
NPS (Net Promoter Score),
10–11, 69, 120
number of emails worked per
period, 29, 120

office administration, 121
Omidyar, Pierre, 9
OneDrive, 31, 120–121
one-off sales, 52, 121
OnPay, 28
OpenAI, 94–95, 99, 121
Operational Excellence (OpEx),
xiii, 4–5. *See also* continuous
improvement
data analytics for, 52–54

defined, 121
getting started, *54*, 54–55
key indicators of, 52
measuring, 52
spreadsheets for, 53–54
summary of, 55
Sundstrand's need
for, 49–52, 55
Oracle, 28
org chart, 121
outbound calls, 17, 121
outsourced billing service, 121

pain points, 58–64, 86, 121
Panetta, Kasey: *Gartner Top
Strategic Technology Trends
for 2021*, x
Patene, Mark: *How to Downsize
Your Business Successfully*, 79
Pavlina, Steve, 57
Paylocity, 28
PayPal, x–xi, 6
ACH transactions by, 37
Bill Me Later, Inc.
acquired by, 88
Customer Relationship
Management's use of, 22
as an eBay subsidiary, 37–38
Identity and Access
Management software
used by, 12–13
Multi-User Experience
with, 38–39
recruitment at, 78–79

regulatory agencies'
    focus on, 38–39
  reporting processes of, 38–39
  telephone contact
    tools of, 10–11
payroll, 28, 31, 121
PBX (Private Branch
  Exchange), 13, 121
Pelrine, Kellin, 93–94
personal computers' importance to
  businesses, 97
personal identifiable information
  (PII), 97, 121
per unit of activity, 41, 43, 121
pilots, 32, 61, 63, 121
POC (proof-of-concept),
  42, 62, 122
point-of-sale (POS)
  system, 21, 121
predictive power, 121
pricing models, 19–20, 44, 121
Private Branch Exchange (PBX),
  13–14, 121
process bugs, 122
process consultants, 122
process improvement, 122
product bugs, 79, 122
product development, 81–82, 122
proof-of-concept (POC),
  42, 62, 122
proposal for service vs. Request for
  Proposal, 44–45
published list price, 44, 122
purchase-to-delivery lead
  time, 68, 122

Qaultrics, 23
*Quality, Productivity and
  Competitive Position*
  (Deming), 59

raw material inventory, 122
Ready, Bill, 37
recruiting process, 122
reporting or dashboarding
  segment, 42, 122
reporting software, 17–18
Request for Proposal (RFP), xii,
  40–42, 43, 44–47, 119, 122
revenue, 68, 122
RFP. *See* Request for Proposal
Rogers, Will, 113
Ruby, 20

SaaS application suite, 122
Salesforce, 21–22, 99, 122
SAP, 28
schedule models, 31, 122
Schulman, Dan, 37
Schulman, John, 94
search engines, 94
Securities and Exchange
  Commission (SEC),
  99–100, 123
Sedol, Lee, 93
segmentation, 41, 123
selection process, 40–41, 123
self-service, 13–14, 19,
  23, 59, 123
service culture, 123

Index | 135

service pricing, 68, 123
Shedd, John Augustus, 91
Siebel, Tom, 46
Siebel CRM, 46–47
Silicon Valley startups, 52
sourcing decisions, 75, 123
SOW (Statement of Work), 45, 123
staffing, xiii
   automation's role in, 74
   and business culture, 76
   cost of onboarding, 74
   during downsizing, 79
   and leadership, 76–78, 80
   of Maggie's nursing business (case study), 73–74
   networking's role in, 74–75
   recruitment, 78–80
   single key employee vs. multiple cross-trained employees, 75–76
   summary of, 80
   tenured staff, 45, 123
   and timely decisions for growth, 74–79
stakeholders, 40–41, 59, 63, 123
Statement of Work (SOW), 45, 123
Stonecipher, Harry, 49–50, 55
Sundstrand, 49–52, 55
Sutskever, Ilya, 94

technologies as part of Total Experience, xii
   automation, 12–13
   eBay's use of, 9–13, 69
   email, 9–11, 13
   Identity and Access Management software, 12–13
   Interactive Voice Response systems, 13–14, 23, 59, 116
   invoicing/bookkeeping platforms, 13
   licensing of, 18–19
   for password resets, 12–13
   PayPal's telephone contact tools, 10–11
   Private Branch Exchange platforms, 13–14
   summary of, 15
tenured staff, 45, 123
Tesla, 6–7, 23
Total Experience. *See also* future of Total Experience; getting started in Total Experience; mistakes to avoid
   applicability to businesses, xiv, 7–8, 13–14
   components of (*see* Customer Experience; Employee Experience; Multi-User Experience; Operational Excellence)
   cost input and benefit balance of, 5–6
   defined, 5, 123
   importance of embracing, ix–xiv
   for nonprofits, 7

origins of the methodology, 3, 6–7
relevance to your business, 113
summary of, 8
Total Quality Management (TQM), 81–82, 123
Toyota, 54, 81
TQM (Total Quality Management), 81–82, 123
transaction fees, 123
Twain, Mark, 57

UKG, 28
U.S. Leather, x

virtual reality (VR), 109–110, 123
Visa, 37
visualization platforms, 118, 120, 124
VR (virtual reality), 109, 123

waiting lists, 84, 86, 96, 124
web chat, 23, 32
weighting and scoring method, 124

Welch, Jack, 49
Wells Fargo, 38
WFM. *See* Workforce Management applications
white-glove service, 20–21, 124
Whitman, Meg, 10
Workday, 28
workflow applications, 12, 124
Workforce Management (WFM) applications, xii
defined, 124
for Employee Experience (EX), 29–34
purchasing, 33–34
workforce planning, 62, 124
work schedules, 30, 124

X.com (formally Twitter.com), 96

Y2K (year 2000), 45–46, 124
year-over-year revenue growth, 67, 124

Zaremba, Wojciech, 94

ACKNOWLEDGMENTS

I want to thank my wife and family for their support and patience. Likewise, I have received the support of numerous friends and colleagues who have acted as editors for several of the stories represented. For those inspirational leaders highlighted, I wish to extend a sincere thank you. The learnings I have been able to accumulate are attributable to the opportunities that you afforded me. Also, to my book design, editing, and marketing team – you are simply outstanding. Lastly, I must give thanks to Gartner Inc.'s thought leaders. They crystallized the various experience disciplines into an all-encompassing Total Experience methodology. Well done!

Let's Connect!

Thank you for embarking on your Total Experience journey and for investing your time in reading my book! Your commitment to reaching this point is a testament to your dedication to your business and your ongoing pursuit of knowledge. I hope you found this book to be both helpful and inspiring.

Should you require additional assistance with specific challenges, would like to discuss scheduling me for a speaking engagement, or wish to explore bulk purchases of this book with a discount, please scan the QR code below to reach out to me directly. Your satisfaction and success are paramount, and I'm here to support you on your journey.

Before you depart, may I request a favor?

Join me in championing the Total Experience methodology by **leaving an honest review of this book on Amazon**. If you believe this book has made a meaningful impact on you or your business, please consider sharing your success story with others. Your support helps spread the message further. Thank you!

Dar Andrews

ABOUT THE AUTHOR

 **DAR ANDREWS** is a native of Nebraska. He grew up working in a family stable and aiding local farmers. After studying at the University of Kansas and the University of Cambridge, England, he graduated with a Bachelor of Science degree from the University of Nebraska–Lincoln. His plus-thirty-year career spans leadership roles with eBay, Gateway Inc., Home Instead Inc., PayPal, Sundstrand Aerospace and U.S. Leather. He enjoys a background in both Finance and Information Technology.

Dar currently enjoys living on a thirty-acre farm in eastern Nebraska with his wife and twin daughters. He is active in various consulting and community endeavors. Hobby farming, outdoor sports, and mentoring remain his passions.

Contact Dar at dar.andrews402@gmail.com or www.linkedin.com/in/darandrews.

Made in United States
Orlando, FL
08 September 2024